8th National Scientific Conference on Advances in Electrotechnology (Postepy w Elektrotechnologii)

IOP Conference Series: Materials Science and Engineering
Volume 113

Jamrozowa Polana, Poland
23 – 25 September 2015

Editors:

Bronislaw Szubzda
Maciej Jaroszewski
Sylwia Talik
Grzegorz Pasciak

ISBN: 978-1-5108-2229-0
ISSN: 1757-8981

Printed from e-media with permission by:

Curran Associates, Inc.
57 Morehouse Lane
Red Hook, NY 12571

CURRAN ASSOCIATES INC. proceedings .com

Some format issues inherent in the e-media version may also appear in this print version.

Printed by Curran Associates, Inc. (2016)

Institute of Physics
Dirac House, Temple Back
Bristol BS1 6BE UK

Phone: 44 1 17 929 7481
Fax: 44 1 17 920 0979

techtracking@iop.org

Additional copies of this publication are available from:

Curran Associates, Inc.
57 Morehouse Lane
Red Hook, NY 12571 USA
Phone: 845-758-0400
Fax: 845-758-2633
Email: curran@proceedings.com
Web: www.proceedings.com

Table of contents

Volume 113

2016

8th National Scientific Conference Advances in Electrotechnology (Postepy w Elektrotechnologii)
23–25 September 2015, Jamrozowa Polana, Poland

Accepted papers received: 25 January 2016
Published online: 11 March 2016

Preface

011001
OPEN ACCESS
PWE: 8th National Scientific Conference Advances in Electrotechnology (Postępy w Elektrotechnologii)

011002
OPEN ACCESS
Technical Editors

011003
OPEN ACCESS
Conference Scientific Committee

011004
OPEN ACCESS
Competition for Young Scientists

011005
OPEN ACCESS
Peer review statement

Papers

DIELECTRIC MATERIALS

012001
OPEN ACCESS
Dielectric response change of pressboard immersed with mineral oil after replacing insulating liquid with synthetic ester

P. Fatyga pg. 1

012002
OPEN ACCESS
Dielectric properties of polymer composites with the addition of ferrite nanoparticles

A Kisiel, M Konieczny and M Zabska pg. 6

012003
OPEN ACCESS
Analysis of the possibilities of using dielectric foam in the construction of composite high voltage post-insulators

T Mączka, G Paściak, A Jarski and M Piątek pg. 14

012004
OPEN ACCESS
Optimization of electret film forming method

B Łowkis and J Kupracz pg. 24

012005
OPEN ACCESS
The behaviour of water droplets on the silicone rubber surface in an electric field

W Bretuj and A Pelesz pg. 29

PLASMA TECHNOLOGY

012006
OPEN ACCESS
Microplasma technology for carbon micro- and nanostructures formation in gas at atmospheric pressure

A T Sobczyk, A Jaworek and M Klein pg. 36

012007
OPEN ACCESS
Thin-film $TiPbO_3$ varistors obtained by two-source magnetron sputtering

J Ziaja and M Lewandowski pg. 41

012008
OPEN ACCESS
Equivalent circuit of a coaxial-line-based nozzleless microwave 915 MHz plasma source

R Miotk, M Jasiński and J Mizeraczyk pg. 45

012009
OPEN ACCESS
Magnetic thin film deposition with pulsed magnetron sputtering: deposition rate and film thickness distribution

M Ozimek, W Wilczyński and B Szubzda pg. 50

012010
OPEN ACCESS
EMI shielding using composite materials with two sources magnetron sputtering

J Ziaja, M Jaroszewski and M Lewandowski pg. 56

SPECIAL MATERIALS & TECHNOLOGIES

012011
OPEN ACCESS
Design, fabrication and characterization of LTCC-based electromagnetic microgenerators

M Gierczak, P Markowski and A Dziedzic pg. 60

012012
OPEN ACCESS
Evaluation of Polyesterimide Nanocomposites Using Methods of Thermal Analysis

B Gornicka, L Gorecki, K Gryzlo, D Kaczmarek and D Wojcieszak pg. 65

012013
OPEN ACCESS
Study of electrokinetic properties of nanofillers used for the preparation of nanocomposites

B Górnicka, K Gryzło and L Górecki pg. 69

012014
OPEN ACCESS
The research of ceramic materials for applications in the glass industry including microwave heating techniques

K. Kogut, K. Kasprzyk, B. Zboromirska-Wnukiewicz and T. Ruziewicz pg. 74

012015
OPEN ACCESS
Electrical properties of varistors based on tin oxide

W Mielcarek, K Prociów and J Warycha pg. 82

012016
OPEN ACCESS
The investigation of the light radiation caused polyethylene based materials deterioration by means of atomic force microscopy

A Sikora, A Grabarek, L Moroń, M Wałecki and P Kryla pg. 86

012017
OPEN ACCESS
Corrosion resistance of the AISI 304, 316 and 321 stainless steel surfaces modified by laser

B. Szubzda, A. Antończak, P. Kozioł, Ł. Łazarek, B. Stępak, K. Łęcka, A. Szmaja and M. Ozimek pg. 92

012018
OPEN ACCESS
Applying of non-toxic oxide alloys and hybrid polianiline compounds as anticorrosive pigments in organic epoxy coatings

W Szymański, A Halama and J Madaliński pg. 100

MAGNETIC & SUPERCONDUCTING MATERIALS

012019
OPEN ACCESS
Analysis of electric current flow through the HTc multilayered superconductors

J Sosnowski pg. 104

012020
OPEN ACCESS
Acoustic emission during fracture of ceramic superconducting materials

L Woźny, A Kisiel and K Łysy pg. 108

012021
OPEN ACCESS
The interaction between the permanent magnet and ceramic superconductor with organic filler

L Woźny, A Kisiel and A Garbera pg. 114

NOVEL SOURCES & ENERGY STORAGES

012022
OPEN ACCESS
Direct waste heat recovery via thermoelectric materials - chosen issues of the thermodynamic description

Piotr Kolasiński and Ewa Kolasińska pg. 118

012023
OPEN ACCESS
Polymer Materials for the Heat Recovery

E Kolasińska, P Kolasiński and B Mazurek pg. 123

012024
OPEN ACCESS
Energy harvesting in high voltage measuring techniques

Pawel Żyłka and Marcin Doliński pg. 128

IOP Conf. Series: Materials Science and Engineering **113** (2016) 011001 doi:10.1088/1757-899X/113/1/011001

8th National Scientific Conference
Advances in Electrotechnology
Postępy w Elektrotechnologii

The 8th Advances in Electrotechnology Conference was held in Jamrozowa Polana - Lower Silesia Poland.

The Conference was organized jointly by:

- Wroclaw Division of Electrotechnical Institute,
- Wroclaw University of Technology - Department of Electrical Engineering Fundamentals and Electrotechnology.

With the co-organizers' support of:

- Polish Committee of Electrical Engineering - Polish Academy of Science,
- Wroclaw Branch of the Association of Polish Association of Electrical Engineers,
- Polish Committee of Electrotechnical Materials of Polish Association of Electrical Engineers,
- Polish Association for Hydrogen and Fuel Cells.

Science Patron were:

- Polish Academy of Science,
- Polish Committee on Large Electric Systems - CIGRE,
- Polish Association of Electrical Engineers.

The main sponsor of the Conference was the Polish Academy of Science.

Wroclaw Branch of the Association of Polish Electrical Engineers sponsored the *Competition for the Best Presentation* which was organized for young researchers - graduate students.

62 participants arrived at Jamrozowa Polana (near Duszniki Zdrój), which is a famous and well-known centre of winter sports. The conference was attended by representatives (authors) of the following Technical Universities: Poznań, Wroclaw, Łódź, Szczecin, Gdańsk, Warsaw and the Institute of Fluid-Flow Machinery Polish Academy of Science from Gdansk, the Institute of Power Engineering of Warsaw, the Institute of Nonferrous Metals as well as representatives of the main organizers - the Electrotechnical Institute from Warsaw and Wroclaw.

Among the participants there were representatives of the enterprises, i.e. Impact Clean Power Technology S.A, ABB Sp. z o.o. and NTI Pte o.o.

The main topics of the conference were as follows:

- electrotechnical materials and technologies,
- technologies and distributed energy systems,
- electrical energy storages,
- fuel and solar cells,
- nanocomposites - technology and application,
- biomimetic materials,
- plasma technology,
- magnetic and superconducting materials and their application,
- electroconductive polymers,
- surge arresters and varistors.

Head of Organising Committee
Bronisław Szubzda

IOP Conf. Series: Materials Science and Engineering **113** (2016) 011002 doi:10.1088/1757-899X/113/1/011002

Technical Editors

Bronisław Szubzda
Electrotechnical Institute
Division of Electrotechnology and Materials Science,
M. Skłodowskiej-Curie 55/61
50-369 Wrocław
Poland
e-mail: b.szubzda@iel.wroc.pl

Maciej Jaroszewski
Wroclaw University of Technology
Department of Electrical Engineering Fundamentals,
Wybrzeże Stanisława Wyspiańskiego 27
50–370 Wrocław
Poland
e-mail: maciej.jaroszewski@pwr.edu.pl

Sylwia Talik
Electrotechnical Institute
Division of Electrotechnology and Materials Science,
M. Skłodowskiej-Curie 55/61
50-369 Wrocław
Poland
e-mail: s.talik@iel.wroc.pl

Grzegorz Paściak
Electrotechnical Institute
Division of Electrotechnology and Materials Science,
M. Skłodowskiej-Curie 55/61
50-369 Wrocław
Poland
e-mail: g.pasciak@iel.wroc.pl

Conference Scientific Committee

IOP Conf. Series: Materials Science and Engineering **113** (2016) 011004 doi:10.1088/1757-899X/113/1/011004

Competition for Young Scientists

One of the important point during the Conference was competition for young scientists on The Best Presentation of the Research Result.
The famous scientists, from different Polish research centers, were members of Competition Committee.
The Committee awarded 6 prizes and 2 awards:
Awards were founded by SEP (Polish Electrician Association) and PKME (Polish Committee of Electrotechnical Materials)

Competition Committee

Prof. Hanna Mościcka-Grzesiak
Prof. Ryszard Kacprzyk
Prof. Jan Subocz
Prof. Andrzej Sikora
Ryszard Kordas PhD - Head of Committee

Awarded Scientists

1st prize	Daniel Gajda
2nd prize	Radosław Szklarek
3rd prizes	Marcin Lewandowski
	Mariusz Ozimek
	Piotr Przybyłek
Awards	Robert Miotk
	Marcin Lebioda

Peer review statement

All papers published in this volume of *IOP Conference Series: Materials Science and Engineering* have been peer reviewed through processes administered by the proceedings Editors. Reviews were conducted by expert referees to the professional and scientific standards expected of a proceedings journal published by IOP Publishing.

IOP Conf. Series: Materials Science and Engineering **113** (2016) 012001 doi:10.1088/1757-899X/113/1/012001

Dielectric response change of pressboard immersed with mineral oil after replacing insulating liquid with synthetic ester

P. Fatyga, Poznan University of Technology, Institute of Electric Power Engineering, Poland

E-mail: przemyslaw.h.fatyga@doctorate.put.poznan.pl

Abstract The manufacturer of synthetic ester claims that replacing mineral oil with his product does not affect the work of the unit. Despite assurances, this information should be treated cautiously. Insulating liquid replacement, i.e. substitution of oil with ester, and especially intermediate stages of this process can cause problems while evaluating solid insulation moisture of the transformer done by means of the most commonly applied FDS indirect method. The article presents results of model investigations of the dielectric response of pressboard samples immersed with mineral oil, which was replaced with synthetic ester afterwards.

1. Introduction

One of the most important parameters of cellulose insulation of power transformers is its moisture. Moisture increase of insulation causes (among others) decrease of electric strength, initiation or intensification of partial discharges and increase of the risk of the bubble effect [1]. The most accurate and most reliable evaluation method of insulation moisture is the laboratory method of titration (Karl Fischer Titration, KFT). Unfortunately, it requires taking samples of transformer insulation, which most often is not possible. This is the reason why insulation moisture of these devices is evaluated by polarization methods. These include: dielectric spectroscopy in the frequency domain (FDS) [2], dielectric spectroscopy in the time domain (RVM) [3], and the measurement of polarization and depolarization current (PDC) [4].

In Poland evaluation of the transformer's solid insulation moisture is most often carried out using the method of dielectric spectroscopy in the frequency domain. It consists of the measurement of capacity C', C" or dielectric losses tgδ in the frequency range from 10^{-4} to 10^{3} Hz. The measurement results are matched with already possessed pattern characteristics of the dielectric response of model samples, of a known temperature and moisture. When the dielectric response of the model is the same as of the investigated object, it is assumed that their moisture is the same. The FDS method is based on the so-called XY model of insulation (Fig. 1) and a set of pattern characteristics obtained from samples of known cellulose parameters and a dielectric liquid by which it is immersed [5]. The accuracy of the determined insulation moisture depends on a number and quality of possessed moisture patterns.

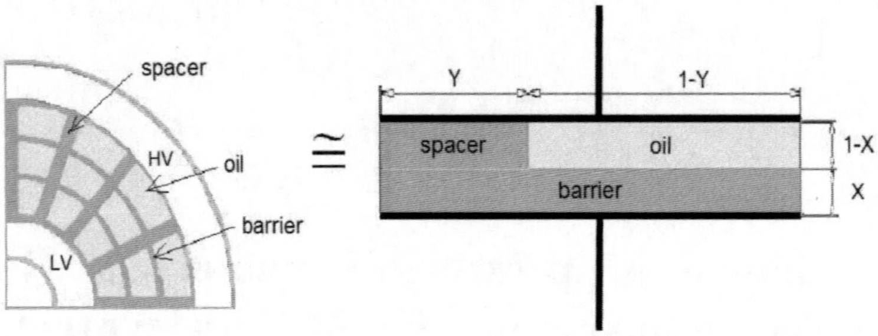

Figure 1. Model XY of the transformer's insulation.

2. Research target

A manufacturer of one of synthetic esters claims that mineral oil can be replaced by their product without any harm to the transformer. Although cases of such an operation are already known, we should, however, analyse a number of effects accompanying the liquid replacement. Both the electroinsulating liquids have different properties, which generates many questions, especially for transformer designers, and to which we do not have a full answer. The measurement reconnaissance done by the author concerned dynamics of replacing mineral oil with synthetic ester process. This evaluation was based on dielectric spectroscopy in the frequency domain.

3. Research object

The research object was a pressboard sample of the moisture of 4.07% and density of 1.2g/cm^3. This sample was initially immersed with mineral oil and next conditioned in a climatic chamber for three days. Moreover, both mineral oil and synthetic ester were used in the experiments. The synthetic ester used for the research was initially dried by covering with it paper rolls which had been dried in a vacuum chamber.

4. Research procedure

The investigated pressboard sample was placed between electrodes in a tight vessel filled with electroinsulating liquid (Fig. 2). This vessel was put in a thermal chamber at the constant temperature of 30°C during the whole period of the investigation (47 days). Between further FDS measurements, the electrodes were moved apart from each other in order to allow easier replacing mineral oil with synthetic ester inside the pressboard sample.

Figure 2. A pressboard sample positioned between the electrodes - view during measurements.

The experiment was carried out according to the following procedure:

- preparation of the pressboard samples and insulating liquids,
- conditioning the samples immersed in oil – 3 days,
- determining the moisture of the pressboard sample using the KFT method,
- placing the pressboard sample at the research set-up,

- investigating the dielectric response of the pressboard immersed in mineral oil (according to the scheme of the measurement system presented in Figure 3),
- replacing mineral oil in vessel with synthetic ester,
- investigating the dielectric response of the sample – repeatedly for a few days.

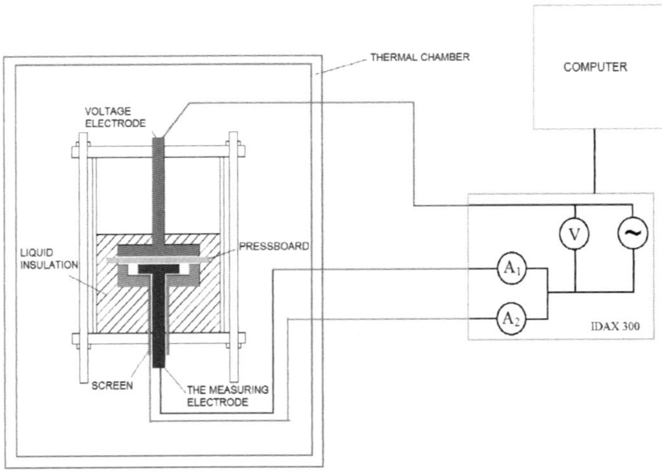

Figure 3. Measurement set-up.

5. Research results

Figure 4 presents obtained characteristics of the dielectric response (ε' i ε") of the insulating system of pressboard and liquid dielectric. These curves were generated on the basis of measurement results and just after changing the insulating liquid from mineral oil to synthetic ester.

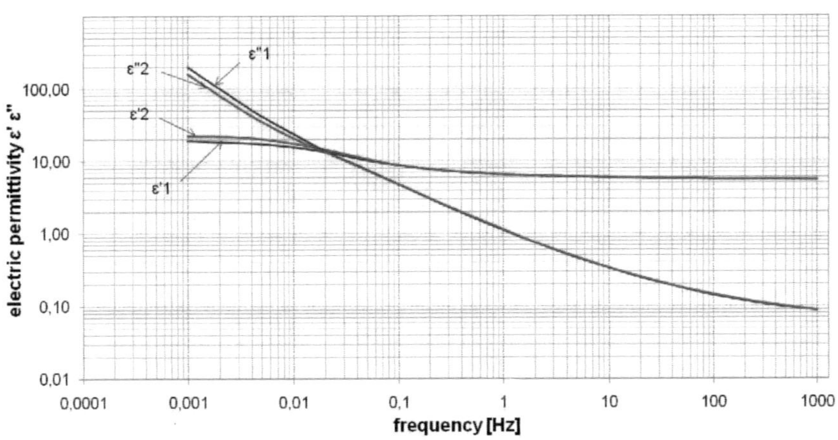

Figure 4. The characteristics of ε ' and ε" pressboard immersed in oil (1) and immediately after changing the oil on synthetic ester (2).

We can see in the graphs that the dielectric response of the pressboard-mineral oil insulating system differs very slightly from the dielectric response of the pressboard-synthetic ester insulating system. It is natural because the synthetic ester had not been able to displace the mineral oil from the pressboard sample in a short time. Noticeable discrepancies in the research results occur only in the range of frequencies from 0.001 to 0.01 Hz, which should be interpreted as a result of the pressboard drying effect by the synthetic ester. Figures 5 and 6 contain graphs of the dielectric response of the paper-synthetic ester insulating system as a function of time which passed from the moment of replacing mineral oil with synthetic ester. We can see in them a successive change of the curves: from a characteristic typical for the pressboard-mineral oil insulating system to a characteristic for the pressboard-synthetic ester insulating system. It results from the figures that after 47 days a full replacement of the dielectric liquid in pressboard still was not obtained. It should be noted, however, that obtaining a model dielectric response of the pressboard-synthetic ester insulating system is practically impossible because there is always a certain residual amount of mineral oil in the measurement vessel.

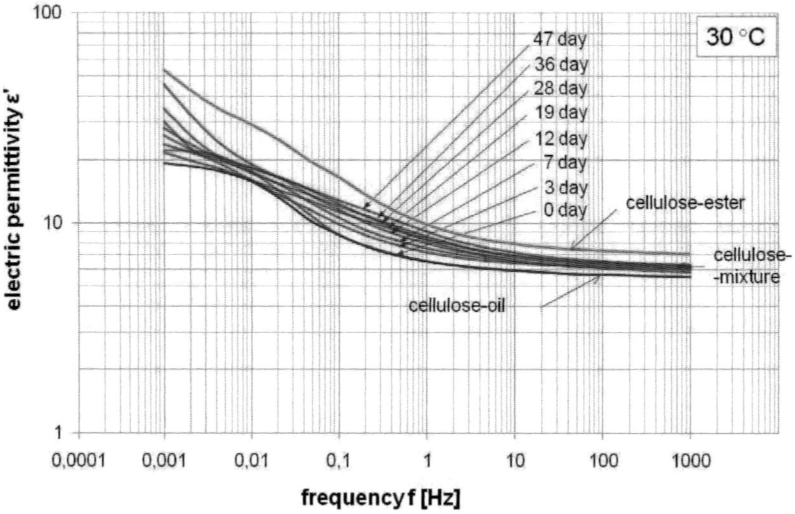

Figure 5. Characteristics ε' as a function of frequency for different times from the moment of replacing mineral oil with synthetic ester.

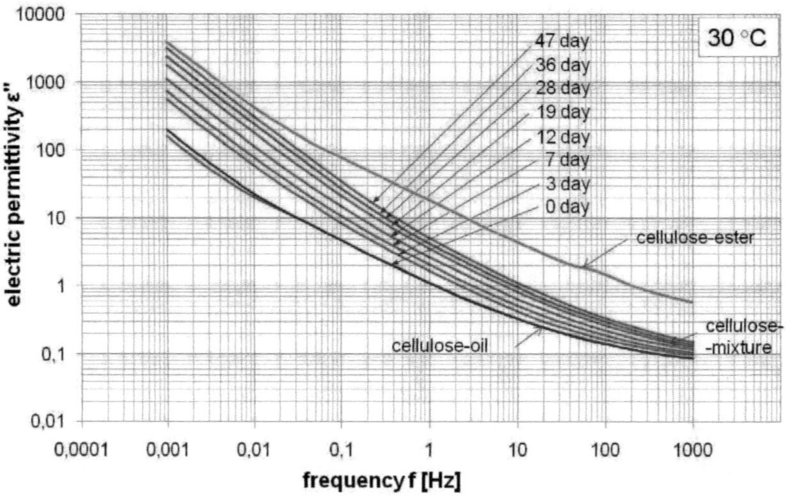

Figure 6. Characteristics ε" as a function of frequency for different times from the moment of replacing mineral oil with synthetic ester.

6. Conclusions

After replacing mineral oil with synthetic ester, characteristics of the dielectric response of the pressboard-dielectric liquid insulating system, taken at intervals, successively shift from a pattern characteristic of the cellulose-mineral oil type to a pattern characteristic of the cellulose-synthetic ester type. Replacing mineral oil with synthetic ester in pressboard is a long-lasting process: after 47 days, complete replacement of mineral oil with synthetic ester was not observed in the investigated sample. A reliable moisture assessment of the cellulose-synthetic ester insulating system after replacing mineral oil with synthetic ester (using the FDS method) is possible either right after the liquid replacement or after a few tens of days.

It should be also included in the further research on results of replacing mineral oil with synthetic ester that there is an effect of cellulose drying by synthetic ester, known from the literature.

7. References

[1] Gielniak J, Moranda H, Neumann S, OssowskI M, Przybyłek P, Walczak K and Mościcka-Grzesiak H Warunki wystąpienia „bubble effect" w transformatorze o izolacji papierowo-olejowej *Konferencja naukowo-techniczna Transformatory w eksploatacji Kołobrzeg-Dźwirzyno 22.04.2005 Energetyka Zeszyt tematyczny nr* ***VI*** 120-122

[2] Bugajny K, Dałek J and Pinkiewicz I 2004 Diagnostyka stanu zawilgocenia izolacji transformatora zmodyfikowaną metodą C_2/C_{50} *Przegląd Elektrotechniczny – Konferencje* nr **1/2004** in Polish 38-41

[3] Dolta B and Shayegani A 2004 Badanie procesów starzenia układów izolacyjnych papier-olej *Przegląd Elektrotechniczny – Konferencje* nr **1'2004** in Polish 54-56

[4] Zengel W S 2003 Applications of Dielectric Spectroscopy in Time and Frequency Domain for HV Power Equipmen *IEEE Electrical Insulation Magazine* **19** 6

[5] Morańda H and KOCH M 2006 Program do analizy odpowiedzi częstotliwościowej (FDS) dielektryków *Przegląd Elektrotechniczny – Konferencje* nr **1/2006** in Polish 162-165

IOP Conf. Series: Materials Science and Engineering **113** (2016) 012002 doi:10.1088/1757-899X/113/1/012002

Dielectric properties of polymer composites with the addition of ferrite nanoparticles

A Kisiel[1], M Konieczny and M Zabska
Wroclaw University of Technology
Department of Electrical Engineering Fundamentals
Wybrzeze Wyspianskiego 27, 50-370 Wroclaw, Poland

E-Mail: anna.kisiel@pwr.edu.pl

Abstract. The aim of the work was examination of the dielectric properties of a new type of polymer nanocomposites based on PVDF (polyvinylidene fluoride), or a copolymer P(VDF-HFP) with addition of ferrite nanoparticles. The addition of nanofillers leads not only to the formation of polar β-phase of PVDF, which shows unique piro-, piezo- and ferroelectric properties used in many applications, but also affects the dielectric and magnetoelectric properties of these nanocomposites. In the work the dielectric properties of polymer composites, such as volume resistivity, permittivity and dielectric loss were investigated Measurements of dielectric parameters were performed in a climate chamber at temperature range of 25-100°C, at selected frequencies in the range 20-200 kHz. The results obtained are valuable not only for an application of this type of nanocomposites in various types of transducers, but also for analysis of the physical phenomena occurring in the polymer composites doped with nanoparticles.

1. Introduction

Composites are materials, which have a main interest of researchers in recent years. They merge at least two components, but usually make them more components resulting in a multiphase material with new, often surprising properties. The ability to combine different types of ceramic compounds, polymer, and metal using a variety of advanced technologies makes creating new composites only limited by imagination of researchers.

As mentioned the composites combine different properties of constituent materials, thus by their appropriate selection and combination unique properties of the obtained composite can be achieved. Among modern composite materials very interesting are piezoelectric polymer composites, especially those. manufactured with PVDF (polyvinylidene fluoride) or its copolymers as the matrix.

This polymer has not only excellent pyro- and piezoelectric properties, but also chemical and mechanical ones. Recently, many of works are concentrated on the PVDF composites with additive of various nanoparticles. Such materials, called nanocomposites can gain very different physical properties than the base material, like composites formed from PVDF and graphene for example. They achieve unique piezoelectric and ferroelectric properties, which make them particularly useful in energy harvesting devices [1]. Compounds of this polymer with aluminum nanoparticles show strong shielding properties of the electromagnetic fields [2]. An addition to the PVDF polymer of carbon nanotubes cause a reduction in resistivity by several orders of magnitude compared to the pure polymer [3].

[1] To whom any correspondence should be addressed.

IOP Conf. Series: Materials Science and Engineering **113** (2016) 012002 doi:10.1088/1757-899X/113/1/012002

The nano-particles used in the PVDF composites cause not only the disclosure of new polymer properties, but also have a significant impact on the process of polymer crystallization. Polyvinylidene fluoride can be found in five polymorphic forms, but only one of the phases - phase β after the polarization process exhibits piezo-, pyro- and ferroelectric properties, therefore in the manufacturing process aims at achieving the maximum content of that phase. Usually, this was achieved by high pressure or stretching of the PVDF polymer under certain conditions. In recent years, novel methods for inducing β phase PVDF are developed. It was observed that the addition of various kinds of nanofilers can enhance that phase formation [4, 5]. Composites based on the PVDF-containing ferrite nanoparticles are particularly interesting. They reveal not only piezo-, pyro- and ferroelectric properties but also magnetic features [6, 7, 8].

Such properties may exhibit material subjected to polarization process but it can be carried out correctly, it is necessary to examine its dielectric parameters. Their values affect the reliability and stability of the manufactured from them sensors, for example. Therefore the work was focused on dielectric properties measurements of a new kind of nanocomposites.

2. Experimental

2.1 Preparation of nanocomposites

All tested samples were manufactured in the Department of Engineering and Polymer Technology of Wroclaw University of Technology. The base polymer of composites was a PVDF poly(vinylidene fluoride) or its copolymer P(VDF-HFP). To the matrix of polymer desired amount of nanoparticles (ferrites and/or montmorillonite particles) was added. The percentage weight of the $CoFe_2O_4$ ferrites varied from 1 to 10% for the samples based on pure PVDF and from 1 to 5% for the P(VDF-HFP) copolymer matrix. Some of the composites were prepared also with addition of 5% of nanoparticles of montmorillonite. This kind of nanofillers after special treatment - the ion-exchange processing, can also influence on formation of PVDF β phase.

After mixing of polymer with the nanofillers samples were extruded and annealed. Manufactured composites had thickness in the range 20-210 μm. For some of them structural testing by X-ray diffraction method was carried out. Results obtained from XRD patterns for doped composites confirmed α to β phase transformation.

In the table 1 and table 2 characterizations of nanocomposites manufactured from pure PVDF or copolymer P(VDF-HFP), respectively, are given.

Table 1. Characterization of nanocomposites based on PVDF.

Sample	**Ferrite $CoFe_2O_4$ content (%)**	**Montmorillonite content (%)**	**Sample thickness (μm)**
P135	1	-	95
P159	-	-	125
ZK5	5	-	20
ZK10	10	-	170
5/BAT5[a]	5	5	125
10/BAT5[a]	10	5	115

Table 2. Characterization of nanocomposites based on P(VDF-HFP).

Sample	**Ferrite $CoFe_2O_4$ content (%)**	**Montmorillonite content (%)**	**Sample thickness (μm)**
P134	5	-	25
P136	1	-	90
P138	-	-	210

2.2 Dielectric properties measurements
Measurements of the dielectric properties of polymer composites containing nanoparticles were carried out in a wide temperature range (25 °C to 100° C) and frequency (20 Hz to 200 kHz). Before measurements aluminum electrodes were deposited by magnetron sputtering onto both sides of composite samples. Measurements of volume resistance were performed for three different voltages, in a two electrodes system. The measuring system consisted of: the climatic chamber Espec model EN-1KPH, gigaohmeter GA-3 (No. 019 / A-7 / 664-1 / T / 1244); multi-function meter HAMEG HM 8118 LRC bridge. Conditioning chamber allowed desired temperature set and humidity control.

3. Results and discussion

In the figures 1-4 dependencies of dielectric parameters (volume resistivity, dielectric constant and dielectric losses factor) on temperature for the selected nanocomposites are presented. From the characteristics $\rho v(T)$ shown in the figure 1 can be seen that for all tested nanocomposites volume resistivity decreases as a function of temperature, but the rate of these changes is not the same. At the low temperature region, to about 50°C, not so significant differences are noticeable, but above that temperature ρv of nanocomposites with ferrites and additionally with montmorillonite nanoparticles decreases rapidly (sample 5BAT5 and 10BAT5).

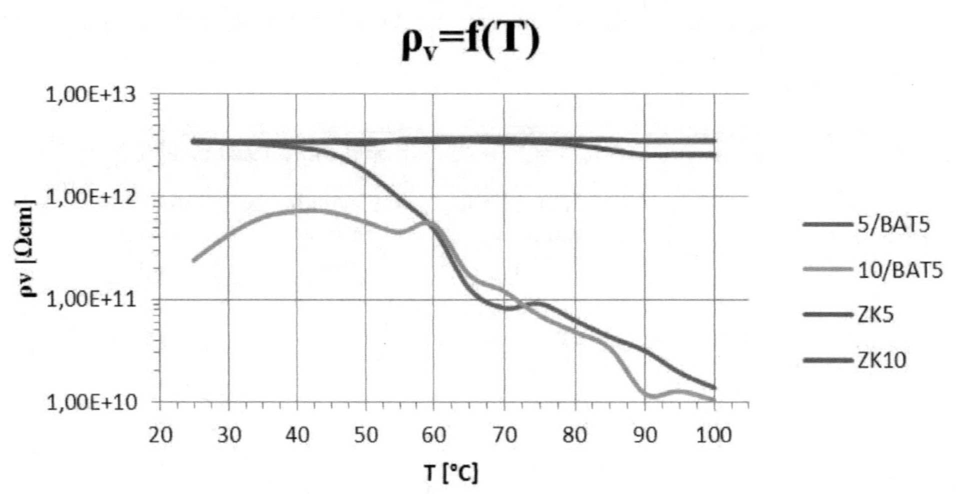

Figure 1. Dependence of the volume resistivity on temperature for the nanocomposites with addition of $CoFe_2O_4$ ferrites (PVDF matrix).

Dependencies of dielectric constant vs. temperature change shown in figure 2 reveal the same behaviour for all presented samples, but obtained values differ significantly.

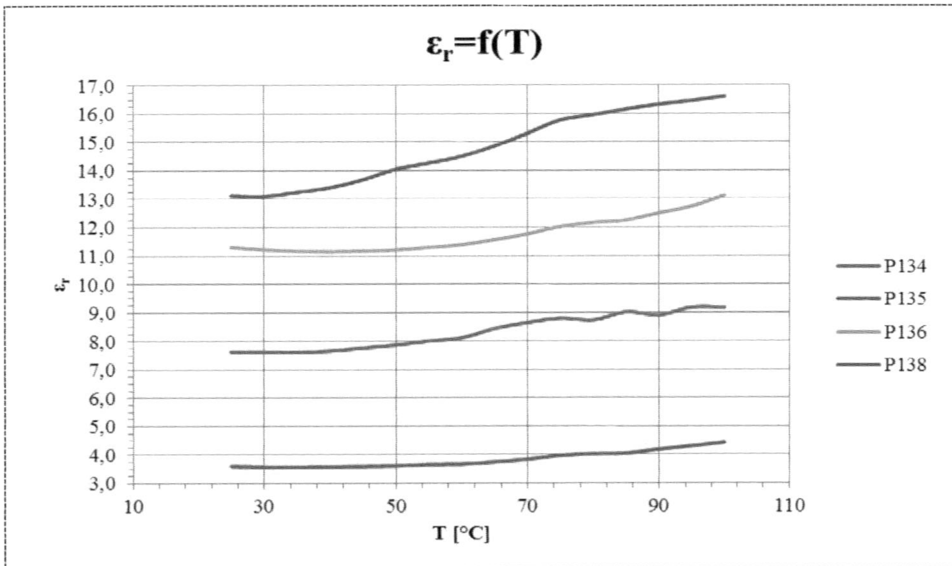

Figure 2. Dielectric constant as a function of temperature (at f=1 kHz) for the nanocomposites with addition of $CoFe_2O_4$ ferrites.

Nanocomposite based on pure PVDF with 1% wt. of ferrites particles (sample P135) shows almost three times lower value of ε than composite with the same ferrites content but manufactured with P(VDF-HFP) copolymer (sample P136).

In the figures 3 and 4 characteristics of dielectric losses with temperature increase are demonstrated. For all composites value of $tg\delta$ increases with the temperature rise. They present the same trend. Only dielectric loss factor for sample without addition of nanoparticles (P138) stay lower that 0.1 in the applied temperature range and shows almost stable value.

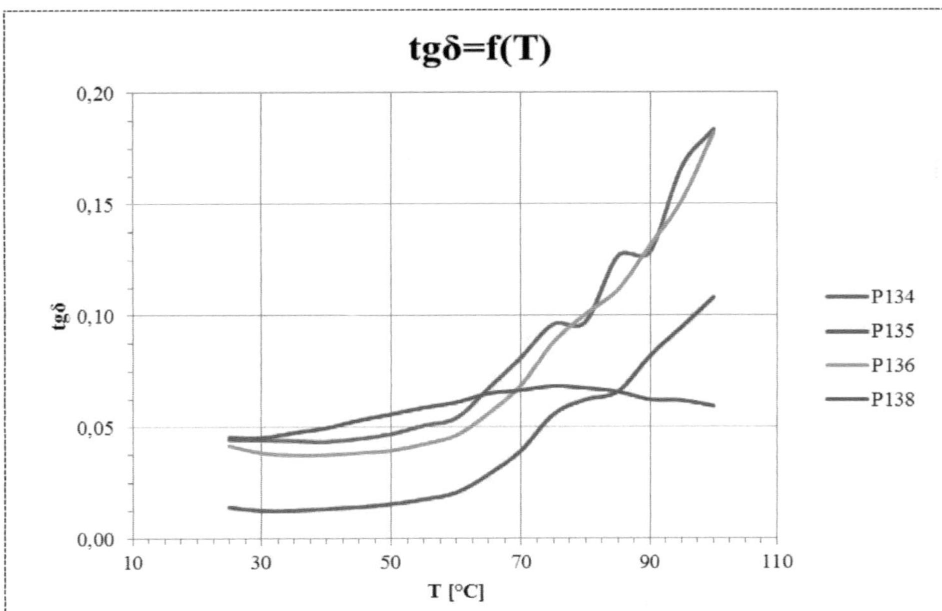

Figure 3. Dielectric losses factor as a function of temperature (at f=1 kHz) for the nanocomposites with addition of $CoFe_2O_4$ ferrites.

Comparing results of *tgδ* obtained for the composites with same content of ferrites nanoparticles, but manufactured with different polymer matrix, can be seen that these PVDF based (figure 4) show lower values of dielectric losses than samples with P(VDF-HFP) copolymer (figure 3).

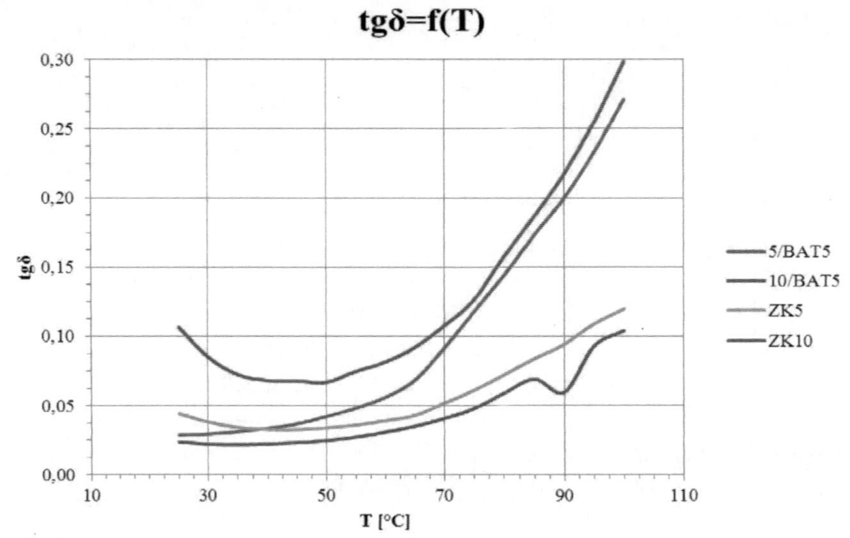

Figure 4. Dielectric losses factor as a function of temperature (at *f*=1 kHz) for the nanocomposites with addition of $CoFe_2O_4$ ferrites (PVDF matrix).

In the figures 5-9 influence of ferrite nanoparticle concentration on dielectric parameters values with temperature as a parameter are presented. Figure 5 and figure 6 demonstrate volume resistivity for composites manufactured with different ferrites weight content for PVDF and PVDF copolymer samples based, respectively.

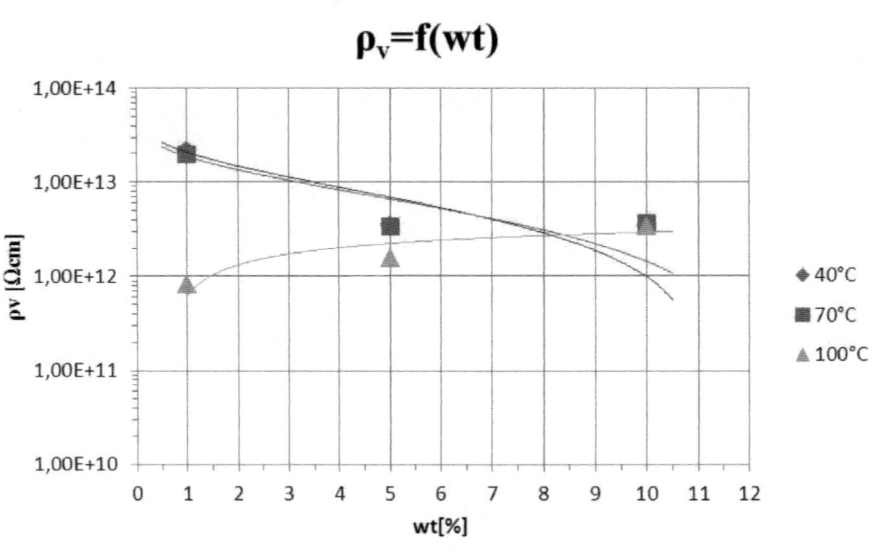

Figure 5. Dependence of the volume resistivity on $CoFe_2O_4$ ferrite nanoparticles weight percentage in composites manufactured with pure PVDF polymer.

IOP Conf. Series: Materials Science and Engineering **113** (2016) 012002 doi:10.1088/1757-899X/113/1/012002

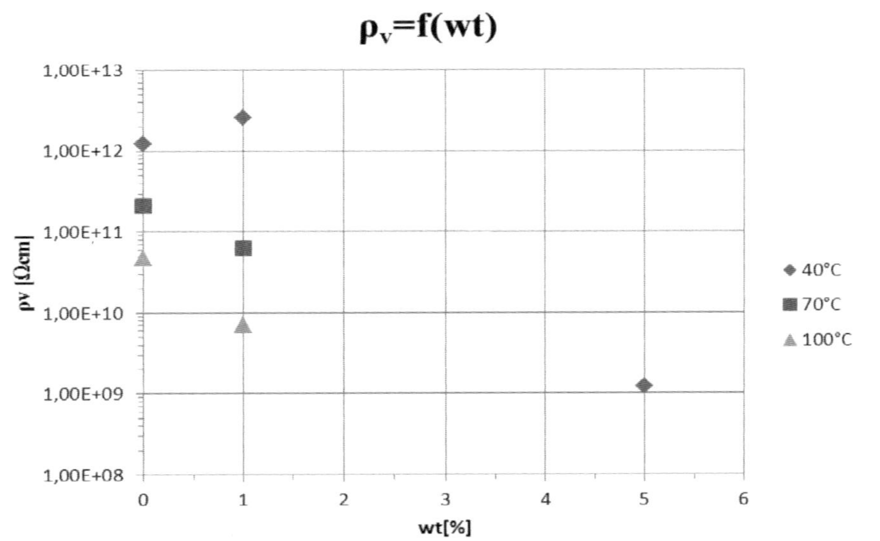

Figure 6. Dependence of the volume resistivity on $CoFe_2O_4$ ferrite nanoparticles weight percentage in composites manufactured with P(VDF-HFP) copolymer.

Dielectric losses vs. ferrites concentration with temperature as a parameter are shown in figures 7 and 8. For composites manufactured with pure PVDF values of $tg\delta$ are lower than for those with P(VDF-HFP) copolymer regardless of nanoparticles content.

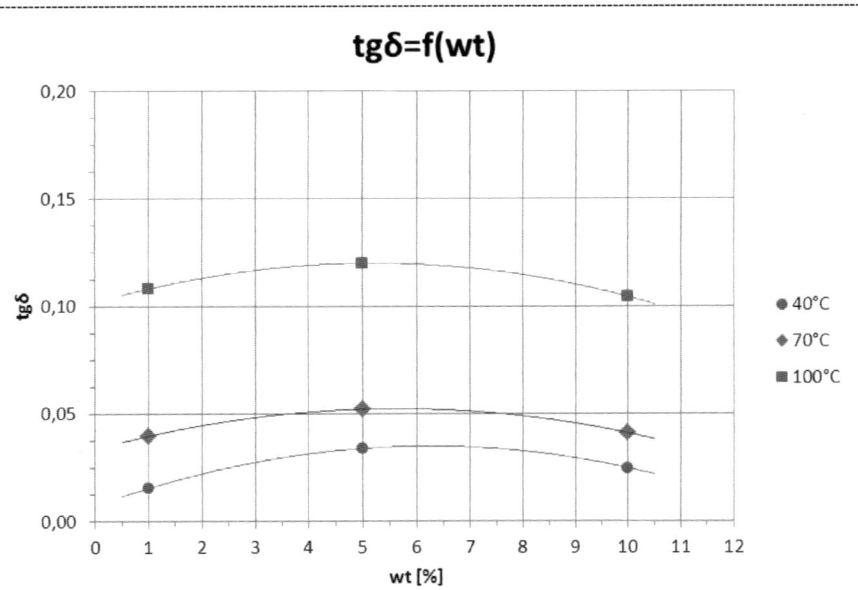

Figure 7. Dielectric losses factor of PVDF based composites as a function of $CoFe_2O_4$ ferrite nanoparticles weight percentage with temperature as a parameter (at f = 1kHz).

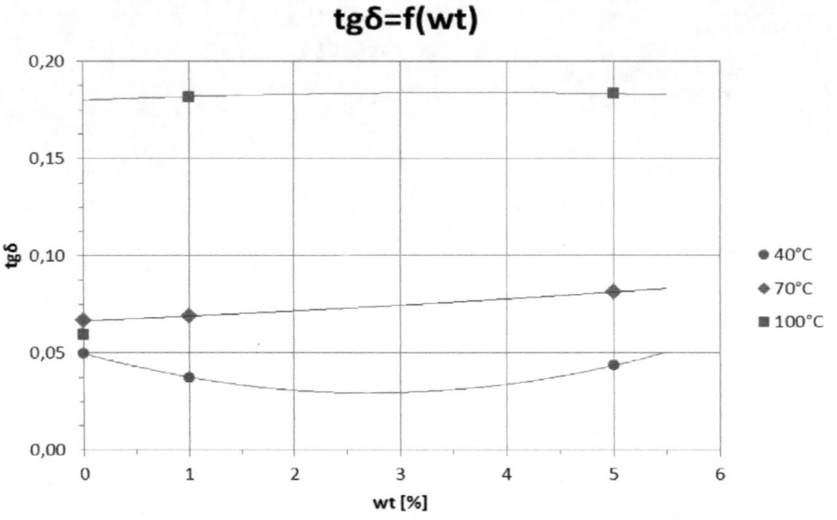

Figure 8. Dielectric losses factor of P(VDF-HFP) copolymer based composites as a function of $CoFe_2O_4$ ferrite nanoparticles weight percentage with temperature as a parameter (at f = 1kHz).

In the figure 9 dielectric constant vs. ferrite nanoparticles concentration is demonstrated for pure PVDF based composites. Obtained results show that 5% concentration of ferrites increases significantly dielectric constant of sample compared to that with 1% content. It changes almost one order of magnitude. High content of nanoparticles (10%) does not change ε value noticeably compared to sample with 5% ferrites concertation.

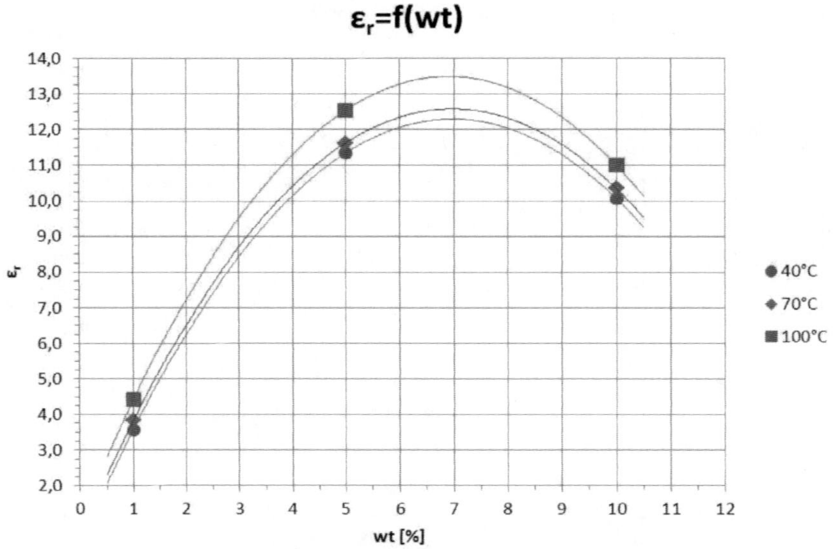

Figure 9. Dielectric constant of PVDF based composites as a function of ferrite nanoparticles weight percentage with temperature as a parameter (at f = 1kHz).

4. Conclusions

In the paper results of dielectric parameters of composites prepared with different $CoFe_2O_4$ ferrite nanoparticles weight content were presented. The obtained results allow drawing following conclusions:

- All measured or determined dielectric parameters (volume resistivity, dielectric constant, dielectric loss factor) change with ferrite nanoparticles addition.
- Changes in dielectric properties are subject to different morphology of composites studied.
- Measured dielectric parameters differ for composites manufactured with pure PVDF and or its copolymer P(VDF-HFP).
- For all nanocomposites volume resistivity decreased with increasing of temperature. This is due to, among other reasons, a reduction of material viscosity and greater mobility of polymer chains, and what goes with it - an increase in the movement of ions.
- More stable values of volume resistivity at temperature change exhibited nanocomposites pure polyvinylidene fluoride (PVDF) based than those with its copolymer.
- The values of dielectric parameters of composites were influenced by the addition of nanoparticles depending on their type (ferrites or ferrites with montmorillonite) and content.
- Increase of nanoparticles caused an increase in dielectric loss factor regardless of the base polymer of composites.

For composites with ferrite nanoparticles addition very important are magnetic properties, therefore research will continue to determine these parameters also in the future experiments.

5. References

[1] Rahman Md A and Chung G S 2013 *Journal of Alloys and Compounds* **581**, 724
[2] Arranz-Andres J, Pulido-Gonzalez N, Fonesca C, Perez E and Cerrada M L 2013 *Materials Chemistry and Physics* **142** 469
[3] Fang-Chyou C 2014 *Materials Chemistry and Physics* **143** 681
[4] Priya L and Jog J 2003 *J. Polym. Sci.* Part B, *Polym. Phys.* **41** 31
[5] Yu L and Cebe P 2009 *J.Polym. Sci.* Part B, *Polym. Phys.* **47** 2520
[6] Martins P, Costa C M, Botelho G, Lanceros-Mendez S, Barandiaran J M and Gutierrez J 2012 *Materials Chemistry and Physics* **131** 698
[7] Cocalves R, Martins P M, Caparros C and others 2013 *Journal of Non-Crystalline Solids* **361** 93
[8] De La Gruz-Montoya E and Rinaldi C 2010 *J. Appl. Phys.* **107** 09B 506

Acknowledgments
This work was funded by the statutory research.

IOP Conf. Series: Materials Science and Engineering **113** (2016) 012003 doi:10.1088/1757-899X/113/1/012003

Analysis of the possibilities of using dielectric foam in the construction of composite high voltage post-insulators

T Mączka[1], G Paściak, A Jarski, M Piątek,
Electrotechnical Institute Division of Electrotechnology and Materials Science
M. Sklodowskiej-Curie 55/61
50-369 Wroclaw, POLAND

E-mail: t.maczka@iel.wroc.pl

Abstract. This paper presents the construction and basic performance parameters of the innovative tubular construction of high voltage composite insulator filled with the lightweight foamed electroinsulating material. The possibility of using of the commercially available expanding foams for preparing the lightweight foamed dielectric materials was analysed. The expanding foams of silicone RTV and compositions based on epoxy resin and LSR silicone were taken into account. The lightweight foamed dielectric materials were prepared according to the own foaming technology. In this work the experimental results on the use of the selected foams for the preparing of the lightweight filling materials to the tubular structure of composite insulator of 110 kV are presented.

1. Introduction

For years the market of high voltage post- insulators line and traction (outdoor and indoor) is entirely occupied by ceramic insulators. They are characteristic by the high specific weight (approx. 2500 kg/m^3) of the porcelain cores (rods) and sheaths of the insulator, and by the need of use of massive mounting and protective hardware. In addition, ceramic insulators are very sensitive to operating conditions, in particular they are low resistant to mechanical impact and thermal shocks that occur, eg. during surface discharge, as a consequence of short circuits and/or surges in the power system [1,2].
There are also composite high voltage post-insulators composed of a fully filled cylindrical core made of mineral (sometimes organic) fibers glued together with the use of an organic binder (typically epoxy resin). In this case on the core, whose function is to transfer various kinds of mechanical loads (bending, stretching, torsion, compression), the sheath of an elastomeric silicone material is applied.
According to the authors [1-4], composite insulators with the outer shield made of silicone elastomer is a better option compared to porcelain insulators because they are more resistant to dirt (high hydrophobicity and the ability to self-cleaning reduces the need for periodic cleaning of insulators). The risk of mechanical damage of composite insulator during transport, assembly and maintenance as well as in the case of emergency (thermal effect of electric arc) is also reduced. Composite insulators with fully filled core have slightly lower weight compared to their counterparts ceramic.
The disadvantages of high voltage composite insulators include the lack of rigidity and a relatively large deflection during application of bending and/or compressive stress (insulators bend and reshape). They also undergo significant angle distortion (twisting) under a load of torque. Better stiffness, relatively low deflection and high torsional endurance have a composite high voltage post-insulators of tubular structure (composite hollow insulators).

[1] t.maczka@iel.wroc.pl.

Insulators of tubular construction with a hollow core are suitable, for example for indoor use, where there are no sudden changes in temperature and are maintained above the dew point (in order to prevent condensation of water from the air in the closed space of the insulator thus to prevent the emergence of discharge on the inner surface of the tube). Composite tubular insulators are also commonly used as insulators with internal gas insulation and used as pressure lead-in bushings, (traction or transformer) insulators and as pressure protective insulation for high-voltage circuit breakers (GIS / GCB).
There are not known to the authors currently commercially available composite insulators of tubular construction filled with a dielectric of low density (not more than 1100 kg / m3). It was therefore decided to develop a new generation of composite high voltage post-insulators of tubular construction with an inner dielectric light filling whose structure is schematically shown in Figure 1.

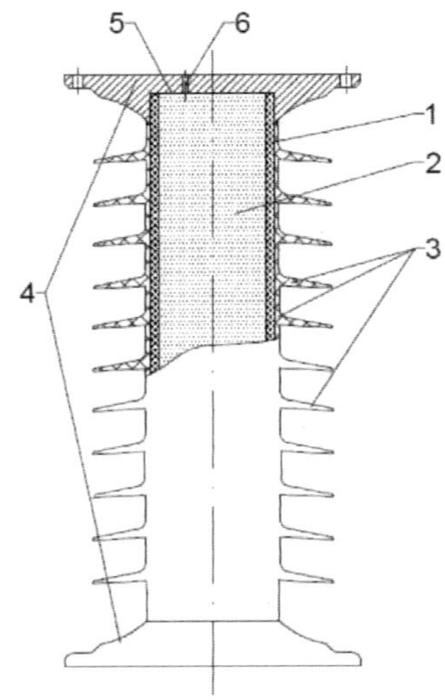

Figure 1. Composite high voltage post-insulators of tubular construction with an inner dielectric light fulfilment
1 – bearing structure (tube) of insulator,2 –light filling with dielectric foam, 3 - elastomeric sheath ,
4 – fittings, 5 – link of glue, 6 – operating hole.

The essence of this design consists in the fact that the interior of the glass-epoxy insulator support tube is tightly filled with expanding foam of dielectric material having a density of 250-500 kg/m3. Due to bearing structure in a form of suitable composite tube these insulators are characterized by high rigidity and good mechanical strength to tensile, bending, compression and torsion. On the other hand elastomeric sheath provides the appropriate creepage, thus and dielectric strength under severe outdoor operating conditions (external) in all environmental zones.
It is believed that a properly made lightweight dielectric filling, eg. foam with high adhesion to both the epoxy tube and metal fittings, fully fulfill the interior of the bearing structure of insulator (tube), thus eliminating the possibility of collection of the moisture inside it and prevent the formation of discharge (creep lines) on the inner surface of the tube.
It is assumed that a composite post insulator with a rated voltage of 110 kV will have the following parameters:

- creepage distance of at least 3813 mm
- striking distance of at least 1075mm,
- rated bending load (SCL) at least 16 kN,
- deflection at a bending force of 4 kN at most 8 mm,

and that they will have to withstand lightning impulse voltage (dry) of more than 575 kV and withstand voltage in rain of more than 275 kV (AC, 50 Hz, effective value).

Preliminary analysis of market showed the economic viability of the planned investment. Developed insulator should ensure trouble-free operation under operating conditions for all environmental zones and possibly optimally meet the (often contradictory) requirements: electrical, mechanical and economical. It is believed that the newly developed product will be 30% lighter than ceramic insulator with the same electromechanical parameters and about twice cheaper. The planned product is to meet outlines and standard requirements for these type of products [5-12].

2. The selected dielectric foams and preliminary research

Choosing the foams for filling the supporting structure of the insulator (tube) you are to take into account material and technological properties such as a high degree of expansion giving the possibility to obtain closed-pore structures with a density of less than 500 kg m 3 and good adhesion to both the fitting and the composite tube core .Not without significance was also uncomplicated way of preparing the composition, the relative low viscosity of components, before and after mixing, the possibility of expansion at room temperature and a short (but not violent) foaming time. What else, material have to be non-combustible or slightly combustible at normal ambient conditions (oxygen index at least 23%) with electrical resistivity of at least 1011 Ωm resistant to temperature changes and thermal shocks -35 / + 50 ($\Delta T \geq 85K$) and of E temperature class of insulation (not lower than 120°C), resistant to moisture diffusion (in case of lack of tightness of the elastomer-fitting-composite triple point or damage of supporting structure of an insulator (a tube). Not without significance is also the availability of material on the market and its price.

Table 1. Some physical properties of dielectric foams selected for testing

Symbol of Material	Type of Material	Characteristics
P1-EPA	Expanding epoxy foam	Bulk density 220 kg/m^3 Oxygen index 21% The resistivity > 10^{13} Ωm, Dielectric loss factor <0.005 Electric strength > 45 kV/cm, The temperature of continuous operation 120°C, Colour: white, Rigid foam, slightly elastic
P2-RTV	Expanding foam based on RTV silicone	Bulk density 300 kg/m^3 Oxygen index 30% Resistivity > 10^{12} Ωm, Dielectric loss factor <0.005 Electric strength > 55 kV / cm, The temperature of continuous operation 140°C, Colour: gray, Elastic foam, flexible
P3-RTV	Expanding foam based on RTV silicone	Bulk density 320 kg /m^3 Oxygen index 35% Resistivity > 10^{13} Ωm, Dielectric loss factor <0.005 Electric strength> 60 kV / cm The temperature of continuous operation 160°C, Colour: graphite, Elastic foam, very flexible
P4-LSR	Foam made of physically foamed LSR silicone	Bulk density 400 kg /m^3 Oxygen index ≤ 35% Resistivity > 10^{14} Ωm, Dielectric loss factor <0.005 Electric strength> 65 kV / cm, The temperature of continuous operation 180°C, Colour: light gray, Elastic foam, flexible

After analyzing of the commercially available materials suitable for the production of dielectric foams, and after an initial determination of their basic material and technological properties the following ones were chosen for testing: a foaming epoxy resin; two types of RTV expansive foams , based on silicone; and the LSR foam obtained with the use of our own method of foaming of silicone. From the selected materials the samples of foams were made using free foaming method under normal ambient conditions. An example of the free foaming of the expanding foam of RTV (P2-RTV) silicone in the glass vessel is shown in Figure 2.

Figure 2. Free foaming of P2-RTV expanding foam.

The foam samples were then tested for their suitability for use as light electrically insulating filling in high voltage post-insulators of tubular construction. The results are given in Table 1.

3. Technological trials to fulfill with foams the full-size models of isolators

In the first stage of the work the tests consisting in choosing optimum process parameters for foaming of material so as to obtain as homogeneous foam as possible with a density of about 200-400 kg / m3, having a closed pore structure and a spherical pores 1-3 mm in diameters.

Trials of foaming were performed within the bearing structure of composite post insulator tube) having an inner diameter of 120 mm and a length of 1200 mm. In order to observe the process of foaming and the expanding of the material and the quality of fulfillment the interior of the tube was closed from both sides with the Plexiglas plates (plates were fasten with metal pins).

The surfaces in contact with the foams were coated with the resolving agent. In the Plexiglas plates some operational openings were made , for filling and venting of expanding composition. Thus prepared closed tubes imitated prototypes of bearing structures of an insulator s and were filled with selected foams.

The process of filling of tubes was performed in the horizontal position using a suitable amount of a composition prepared by a low pressure method. After filling the composition expanded the length of its body thus to form a homogeneous foam filling. In the process of foaming the insulator tube was slowly rotated around the long axis of symmetry making one rotation in approx. 5 sec. When it was found that the foam filled the entire space of the tube, the tube was put into a forced-air drier for a few hours at a temperature of 70°C. An example of the tube after heat conditioning and cooling to room temperature is shown in Figure 3.

Then the supporting insulator tubes (filled with dielectric foams) were cut to a nominal length and their outer surfaces cut to the size of fittings, using a standard technique of cutting. Next the processed and degreased surfaces of the outer ends of the insulator tube and the inner surfaces of the fittings were covered with a thin layer of epoxy adhesive. Then the carrier element of the insulator was push-mounted on -fittings and model of the insulator put into a drier and held at 70°C for about two hours in order to complete fixation of laminated structure.

Figure 3. The supporting insulator tube filled with P3-RTV foam.

For each dielectric foam, one full size model of an insulator (without elastomer sheaths) has been done, a sample of which is shown in Figure 4.

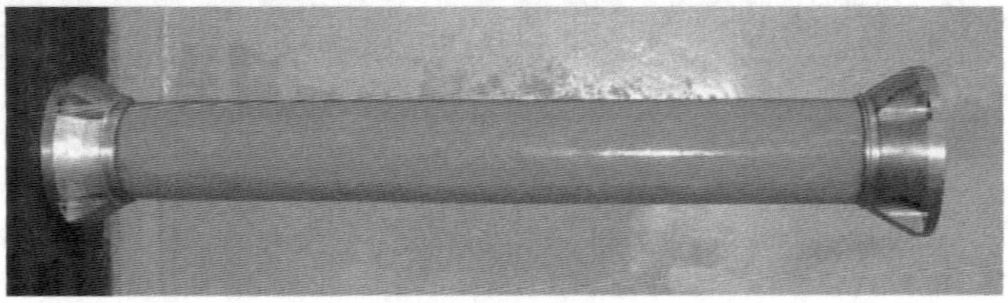

Figure 4. Model of supporting element of composite 110 kV insulator.

After the work proceeded to the end of technological stage the selected tests and checks were performed, in order to evaluate the suitability of the foams as a dielectric light fulfillment for high voltage insulators.

4. Research verification

4.1. Evaluation of the structure of foams and their adhesion to the structure of the tubular insulator
For the morphological evaluation of the foams , **cylindrical** samples, with a height of approx. 40 mm, were cut (with **diamond saw)** from the filled tubes perpendicular to the main axis and subjected to a visual examination.
Figure 5 shows a selection of images of samples taken during the assessment. During visual **examination** the quality of fillings and macroscopic structure of pores was also assessed. **Furthermore** the mechanical strength of foams was examined, especially at composite material–foam junctions. Also foam adhesion to the composite tube was checked, by pressing with a sharp knife.

Figure 5. Images of foam samples taken during the examination of their macroscopic structure and adhesion.

IOP Conf. Series: Materials Science and Engineering **113** (2016) 012003 doi:10.1088/1757-899X/113/1/012003

When assessing the foam structure, the observations were carried out using metallographic microscope Nikon ECLIPSE MA200 using optical measurement program NIS-Elments BR3.10, equipped with camera of DS-FI1 type. Microscopic examination was performed by light reflected from the surface of the foam method, at a magnification of 50x. Figure 6 shows this observation for the P3-RTV foam, taken in the form of a 3D image.
Observations from organoleptic inspection and microscopic inspections of dielectric foams are summarized in Table 2.

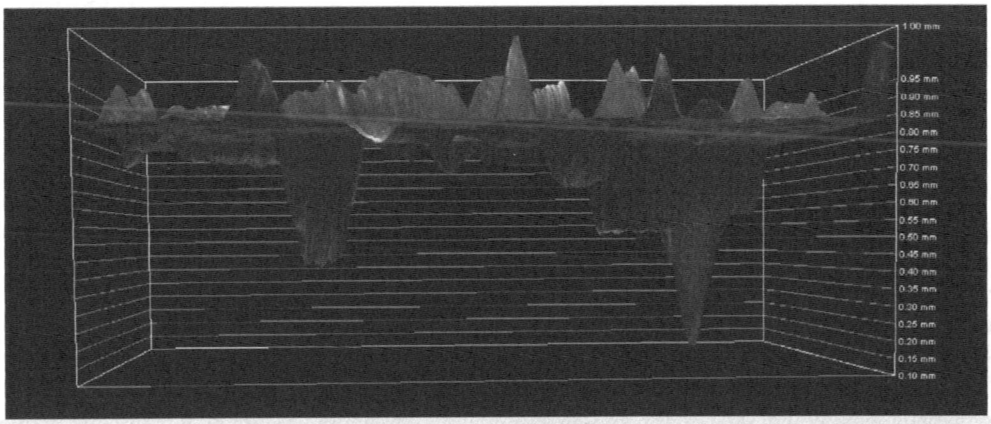

Figure 6. The 3D microscopic image of P3-RTV foam (side view, the scale of 0.10 - 1.00 mm).

Table 2. Morphological evaluation of foams to fill the supporting structure of insulators

Type of foam	Structure	Adhesion	Remarks
P1-EPA Expanding epoxy foam	Hard and rigid foam, rough to the touch, withstands a strong pressing of thumb, quite mechanically resistant. Very good foaming, density 250 kg / m^3, at average. Rather uniform pore structure, most of spherical pores with a diameter less than 1.5 mm. In the cross- break was detected a single spherical cavities in length of 3 mm were detected.	Moderate adhesion to epoxy-glass composite (tubes). Using a sharp edge of the knife it can be easily removed.	Washing the composite before foaming with extraction petrol, and then with acetone improves adhesion.
P2-RTV Expanding foam based on RTV silicone	Foam flexible, soft, with a poor tear strength, under moderate thumb pressure cracks with tending to tear. Generally weak mechanically. Pores structure rather uniform. Well foamed density 320 kg / m^3, at average. Most pores are spherical, less than 1.0 mm in diameter. In the cross-break a few uniformly distributed spherical pores (caverns) 2.5 mm in diameter were spotted.	Very good adhesion to the epoxy-glass composite (tubes). Using a sharp edge of the blade it can be quite easily separated, however it leaves a film on the composite surface, which can be removed only by scraping.	It does not require special preparation of the surface of the composite.
P3-RTV Expanding foam based on RTV silicone	Resilient foam, with moderate tear strength. Under strong thumb pressure splits locally. Generally weak mechanically.	Good adhesion to epoxy-glass composite (tube). Using a sharp edge of the blade it can be quite easily separated. It leaves a film	It does not require special preparation of the surface of the composite.

	Well foamed, with density of about 350 kg / m^3. Uniform pore structure. Most of spherical pores with a diameter of 1 mm.	on the surface, which can be a relatively easily scraped off.	
P4-LSR Foam made of physically foamed LSR silicone	Flexible foam with good tear strength, even when cut. Relatively robust mechanically. Poorly foamed, bulk density 450 kg / m^3. Poorly homogeneous pore structure. Most of the spherical pores having a diameter in the range 0.5-1.5 mm.	Poor adhesion to epoxy-glass composite (tube). Using a sharp edge the knife it can be quite easily removed.	Washing the composite with the use of the so-called 'primer ' designed for LSR silicones significantly increases its adhesion to the composite.

4.2. Dye penetration test
Dye penetration **test fall** within the scope of the design qualification and is an essential structural and material test **when** verifying the quality of core **of** composite **post-**insulators**, line and traction ones [5,14].**
Dye penetration test was carried out on according to section 9.4.1 of PN-EN 62217**:** 2013**.** For the dye penetration **test, the** 10 discs**, 10 mm high,** were cut from the supporting elements of an insulator model so as to the test specimens from them.
The samples were then placed in a jar with steel beads flooded with dye 2-3 mm above the level of the beads. The test relied on visual inspection of the dye penetration through the upper edge of the samples and measuring the penetration of the dye (observation period up to 15 minutes).
Dye penetration was also evaluated at the foam-composite tube contact. During the evaluation of dye penetration some photographic documentation of foams was made. Figure 7 shows the P1-EPA foam images taken during this test.

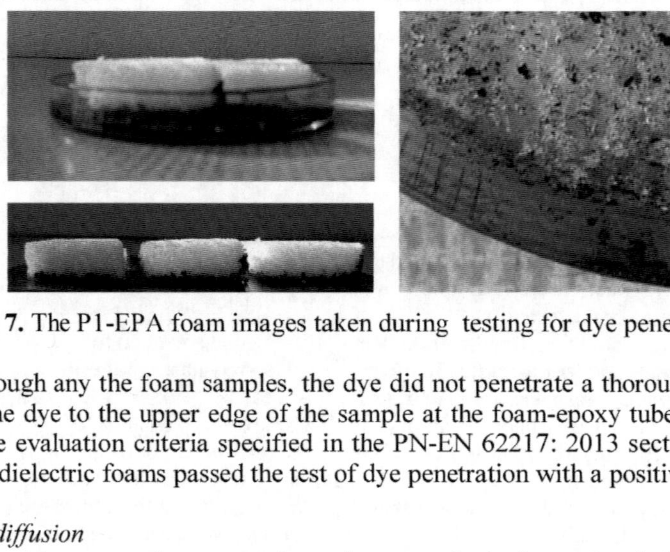

Figure 7. The P1-EPA foam images taken during testing for dye penetration.

During the test, through any the foam samples, the dye did not penetrate a thoroughly, and there was no penetration of the dye to the upper edge of the sample at the foam-epoxy tube interface. Thus, in accordance with the evaluation criteria specified in the PN-EN 62217: 2013 section 9.4.1.2 must be recognized that the dielectric foams passed the test of dye penetration with a positive result.

4.3. Test for water diffusion
Due to the fact that they are not known to the authors standards for testing hollow insulators with filling, and the more with the light fulfillment as a dielectric foam, it was decided to subject the models of isolators to water diffusion test according to the procedure outlined in the standard [6].
Testing for diffusion of water also falls within the scope of the design qualification and is an essential structural and material test when verifying the quality of supporting element of composite insulators designed for lines and tractions [5, 14]. It should be noted that this test is designed for composite insulators with solid core. As for the dye penetration test, the test specimens in a form of 30 mm high discs were made from the supporting element of insulator models. The samples were boiled for 100 h

in deionized water containing 0.1% NaCl (in accordance with the guidelines added in BS EN 62217: 2013 standard, section 9.4.2.3).
Figure 8 shows a sample cut from the insulator model filled with P1-EPA foam, in an early stage of cooking (pre-exposing). As shown, the sample stay afloat (floats) in a volume of water.
Also, samples of filled with foams P2-RTV and P3-RTV "floated" during cooking. This state was observed throughout the whole 100 hour period of exposure. When it comes to P4-LSR foam sample it dropped to the bottom of the vessel and throughout the time of exposure was only slightly raised by bubbles of boiling water. After cooking, the samples were removed from the vessel with boiling water and placed for 15 minutes in another container filled with tap water with an ambient temperature. The voltage test was to be carried out in 3 h after removal of samples from the cooking vessel.

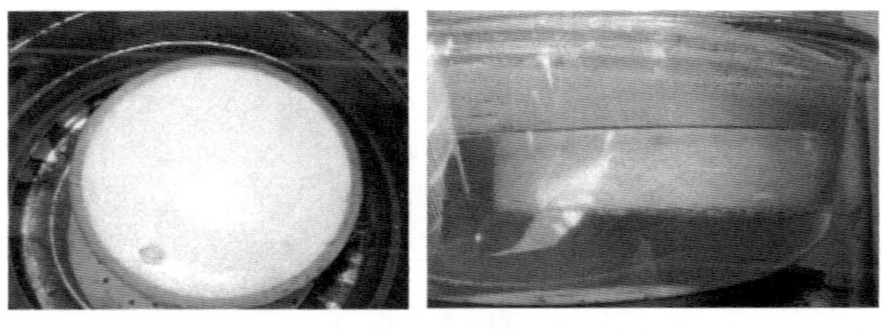

Figure 8. A sample with P1-EPA foam filling, in the early stage of cooking.

After 15 minutes of conditioning in tap water (with an ambient temperature), it was found that the P2-RTV i P3-RTV foam samples contracted and partially detached from the composite tube which illustrate examples of images in Figure 9.

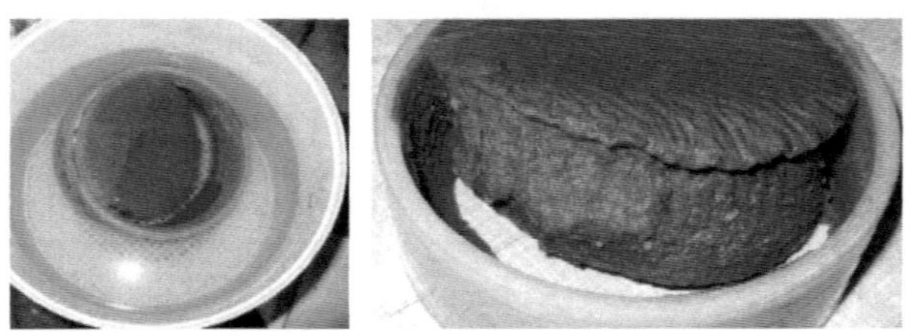

Figure 9. View a sample with P3-RTV foam filling, after test for diffusion of water.

It was also found that the pores in the P2-RTV and P3-RTV foams comprise macroscopic water. In view of these the facts, the verifying voltage test for P2-RTV and P3-RTV samples of foams was pointless.
Samples of the P1-EPA and P4-LSR foams (after exposure by cooking and conditioning in water with an ambient temperature) were designed to verifying voltage test. The scheme and view of measuring system is shown in Figure 10.

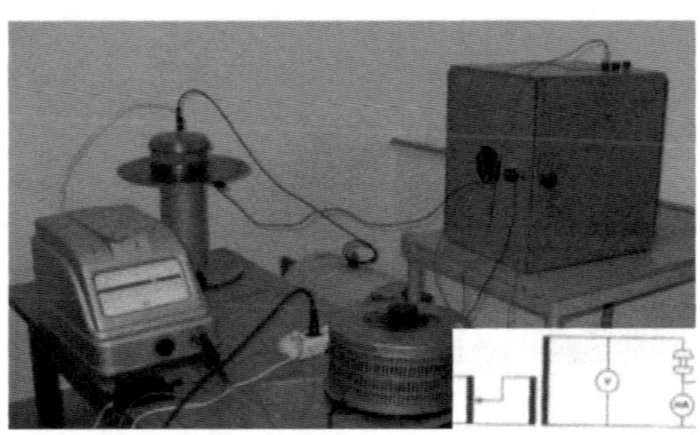

Figure 10. The view and scheme of measuring system used in high-voltage test.

In accordance with the procedure specified in EN 62217: 2013 standard, section 9.4.2.4, directly before voltage trial the surfaces of the samples were dried with filter paper. The samples were then placed between electrodes and test voltage with a value of 12 kV (RMS, AC 50 Hz) was applied. Voltage test time was 60 seconds and was calculated from the moment of achievement of the required testing voltage. According to evaluation criteria (section 9.4.2.5 of the PN-EN 62217: 2013), while applying testing voltage any puncture or surface jump cannot occur. Also during the entire test, the leakage current cannot exceed 1 mA (RMS).

In the case of the sample filled with P1-EPA foam during the test voltage, while the required test voltage was attaining, the throughout puncture occurred , and a visible dark path (charring) appeared on the surface of the foam which shows the photograph in Figure 11.

Figure 11. The view of the dark path (charring) at the surface of P1-EPA foam, after the voltage test.

But for sample with P4-LSR foam filling, while attaining the required value of the test voltage, the current flowing through the sample exceeded 1 mA. After reaching the voltage of 12 kV the current flowing through the sample (leakage current) exceeded 30 mA and the trial was stopped. There was not fired path. Table 3 shows the water diffusion test results for the all tested dielectric foams.

Table 3. Selected properties of dielectric foams chosen to testing

Lp	Type of foam	Leakage current mA	Voltage, kV	Observations	Evaluation
1	P1-EPA	> 1,0	12,0	Through breakdown, visible dark path (charring)	negative
2	P2-RTV	-	-	Breakage of filling, macroscopic water in the pores of the foam	negative
3	P3-RTV	-	-	Breakage of filling, macroscopic water in the pores of the foam	negative
4	P4-LSR	>>1,0	12,0	Excessive leakage current	negative

In accordance with the evaluation criteria set out in the PN-EN 62217: 2013 section 9.4.2.5, none of the tested dielectric foams, intended as light fulfillment to the composite insulators of tube structure, passed the test for diffusion of water with positive result.

5.Summary

As a result of conceptual works, design and technological developments, the novel construction of high-voltage hollow composite insulator has been developed.

As electrical insulation foams to fill the supporting structure of the developed insulator it was decided to use the commercially available materials (epoxy resin foaming, two expansive foams based on RTV silicone and a foam obtained by our own method of foaming LSR silicone). For each of the selected material the technique of filling the interior of the tubular structure of the insulator with the expanding foam was developed.

On samples taken from the full-size models of insulators filled with the produced foams, the penetration of the dye and water diffusion tests, which verify [5, 12] the quality of core of composite line and traction insulators, have been performed. The evaluation found that all of the dielectric foams passed the test for dye penetration positively. In contrast, none of the foams received a positive assessment in testing for water diffusion.

In summary, the carried out research has shown that foams made of the selected materials, despite their high dielectric properties and process parameters should not be used as an electrical insulation for tubular (hollow) insulator. Contraindications to the use is a negative test result for diffusion of water, and this attempt is contained in the standards [5, 6, 12] and falls within the scope of design quality qualification.

Basing on previous experience, the carried out works and technical examinations it is anticipated that the tested materials may be suitable for fulfillment of insulation pipes that should meet the requirements of the standard [13].

As an electrically insulating light fulfillment to insulators of tubular construction suitable are silicone LSR, as shown by the work of the authors at issue in the patent [14] , and relatively new materials which are silicone gels.

6. References

[1] Usal S., Karthika V. P., Edison Selvaraj D., Kiruba Sankar M., Dinesh Aravind C., 2012, *International Journal of Latest Research in Science and Technology*, 1 (3) 247-253

[2] Mazurek B., Mączka T., Gmitrzak A., 2000, *Proceedings of The 6th International Conference on properties and Applications of Dielectric Materials*, Xi'an, China

[3] Chavez J.L.F., *INMR November/Disember* 2003 11 (6) 30-34.

[4] Świerzyna Z., Paściak G., Mazurek B., Olejnik J., 2009, *Materials Science-Poland*, 27, 4/2.

[5] PN-EN 62231:2008, Composite station post insulators for substations with a.c. voltages greater than 1 000 V up to 245 kV – Definitions, test methods and acceptance criteria.

[6] PN-EN 62217:2013-6 Polymeric HV insulators for indoor and outdoor use – General definitions, test method and acceptance criteria.

[7] PN-EN 60071-1: 2008, Insulation co-ordination – Part 1: Definition, principles and rules.

[8] PN-EN 06303: 1998, Exposure of outdoor insulation to pollution and selection of insulators under polluted conditions.

[9] PN-IEC 815: 1998, Guide for the selection of insulators in respect of polluted conditions.

[10] PN-IEC 60273: 2003, characteristics of indoor and outdoor post insulators for systems with nominal voltages greater than 1 000 V.

[11] *Polymer vs porcelain suspension insulator equivalence - what to use,* 2004, http://www.hubbellpowersystems.com EU1205-HR (as at year 2015).

[12] PN-EN 61952:2010 Insulators for overhead lines – Composite line post insulators for A.C. systems with a nominal voltage greater than 1000 V

[13] PN-EN 60855:1999, Insulating foam-filled tubes and solid rods for live working.

[14] P.412376, 2015, *Patent Application*, T. Mączka, G. Paściak, M. Piątek, A. Jarski.

Acknowledgments

Authors gratefully acknowledge the financial support received from: Operational Programme Innovative Economy 2007-2013, Priority 1 Research and development of new technologies, Section 1.4 Support for special projects, POIG.01.04.00-00-150 / 13 "Innovative technology for composite insulators of the Electrotechnical Institute"

Optimization of electret film forming method

B Łowkis, J Kupracz
Wroclaw University of Technology, Department of Electrical Engineering Fundamentals, 50 – 370 Wrocław, Wybrzeże Stanisława Wyspiańskiego 27

E-mail: bozena.lowkis@pwr.edu.pl, jan.kupracz@pwr.edu.pl

Abstract. The aim of this investigation was to develop a method of fabrication of electrets made from PTFE foil of a 0.5 mm thickness, from Ensinger GmbH. Parameters allowing assessment of the electrets in terms of technical applications are: charge lifetime τ and **equivalent** voltage U_z. Measurements of electrets **equivalent** voltage U_z has been carried out using compensation method. Assessment of lifetime required employing a method utilizing thermal stimulation of charge relaxation process. In the investigation, a continuous measurement of **equivalent** voltage in the condition of linear increase in temperature **- TSU$_z$** was realized. It was stated that electret properties of a foil depend on the preparation conditions of the material and the forming temperature. The employed method of lifetime measurement enables verification of the applied procedure.

1. Introduction

For practical applications, electrets from polymeric materials are made by implantation of electric charge from an external source [1]. In the literature, there are described the methods for stability improvement of collected charge by chemical modification of polymer surface before the forming process or its treatment with a low-temperature plasma. Fabrication of electrets at elevated temperatures is also employed. A chemical modification of PTFE foil, 13μm, was reported in [2]. The modification was carried out by using tetraethoxysilane, tetrabutyl titanate and orthophosphoric acid. After modification, the foil was formed in corona discharge in an air triode system. Measurements of thermo-stimulated depolarization were performed at a linearly increasing temperature with a rate of 5°C/min. Chemical modification of the polymer led to an improvement of stability of the collected charge. An improvement of electret charge stability was also obtained by a treatment of the polymer surface with plasma of 20 W [3]. The plasma was developed in O_2/He, O_2 and H_2 atmosphere, at a supply voltage frequency of 13.56 MHz . The samples were formed in an air triode system. In [4], a method of electrets preparation from PTFE foil formed at temperatures of 30°C, 90°C and 150 °C, at a voltage of the corona electrode of 12 kV was presented. The tests of charge stability depending on temperature were performed after five-day storage of electrets in ambient environment. It was stated that the properties of electrets from PTFE foil depend on the forming temperature.
In the paper, the effects of temperature and aging time of samples before the forming process and the temperature of charge implantation on the electrostatic properties of electrets have been presented.

2. The method of electrets forming

The investigations were carried out on samples made from PTFE foil of a 0.5 mm thickness, from Ensinger GmbH. The electrets were formed using corona discharge in an air triode system. A scheme of the set-up for forming the electrets is shown in Fig. 1. A sample with a one-side deposited electrode was placed at a distance of 42 mm from a corona electrode with a radius of 50 μm. A grid was positioned at a distance of 10 mm over the sample surface.

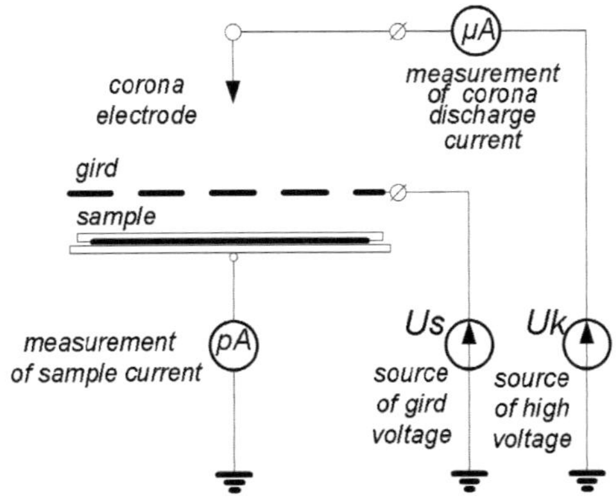

***Figure. 1.** A set-up for electrets forming.*

Due to the need of deposition of an electrode on one of the PTFE surfaces, a chemical modification of this surface has been performed to improve its adhesive properties. The deposited electrodes were characterized by satisfactory adhesion to the substrate, both at the stage of forming at elevated temperature and during the thermo-stimulated tests. The electrode of the sample was grounded through a picoammeter as it has been shown in Fig. 1. The forming of electrets was carried out in air at room temperature (R.T.) T_0 and at temperatures of 100 and 150 °C, at a constant grid voltage Us=1400 V and corona discharge current of 5 μA. Corona discharge voltage was set as +13kV. Before the forming process, the samples were aged for one or four hours. Temperature diagram of the sample before and during the forming process of the electrets at 100°C and 150°C is shown in Fig. 2. The sample was placed on the measuring setup where it was heated up to an assumed temperature for 30 min, and then annealed at this temperature for one or four hours (section $t_1 - t_2$). After , the process of forming started, which was then continued until cooling of the sample. For all temperatures the same geometrical parameters of the measuring stand were maintained.

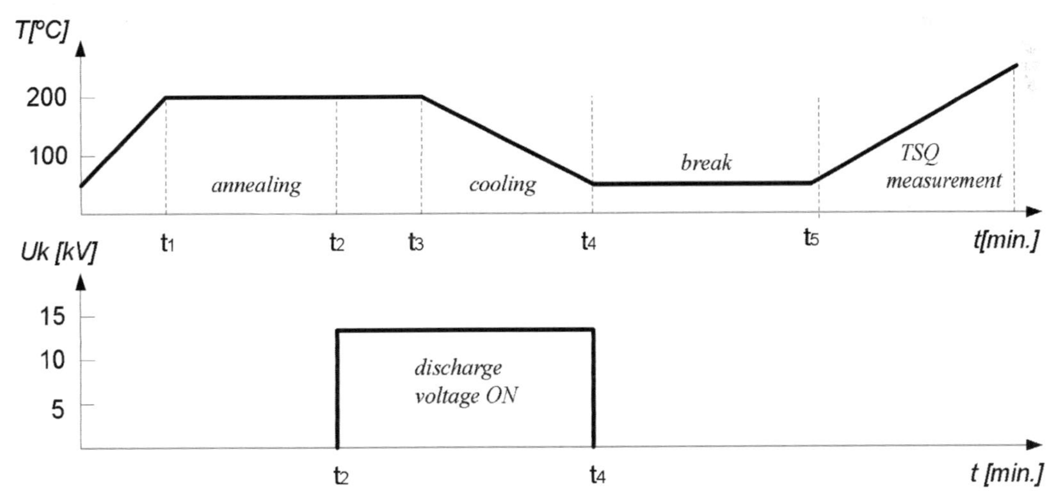

Figure 2. Diagram of temperature change before and during the forming process of electrets versus time.

3.Method of assessment of electret lifetime

Assessment of lifetime of charge τ requires determining the dependence of electrets equivalent voltage at linearly increasing temperature [5]. An exemplary characteristics of thermo-stimulated equivalent voltage of electret with indicates characteristic parameters, necessary for lifetime assessment is shown in Fig. 3.

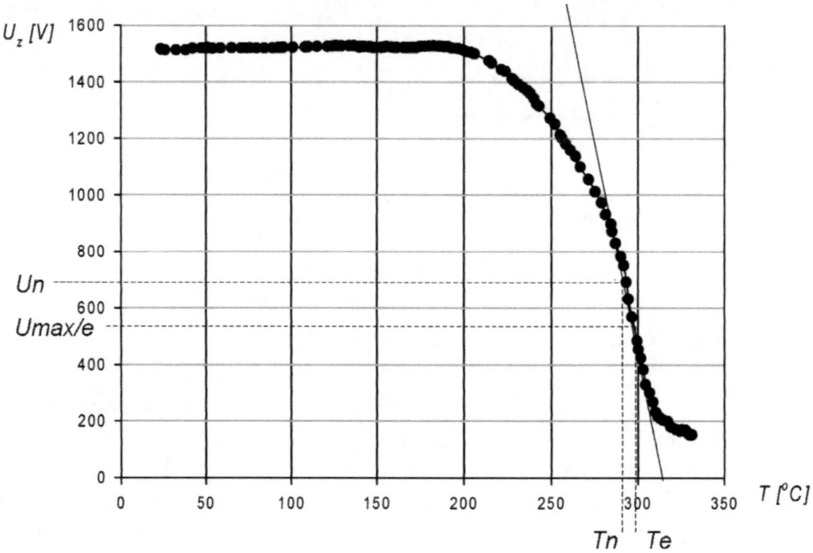

Figure 3. Characteristics of charge decay in PTFE electrets versus temperature.

The run of change in the charge (equivalent voltage) can be described by the Turnhout equation [6]:

$$q_{SA}^{T(t)}(T) \cong q_{SO} \exp\left[-\frac{kT^2}{bW\tau_0}\exp\left(-\frac{W}{kT}\right)\right] \tag{1}$$

where:

q_{SA} – effective density of surface charge,
q_{so} – initial value of homocharge,
b – rate of temperature change during investigation,
W – activation energy,
k– Boltzmann constant ($1.38\text{x}10^{-23}$ J/K),
τ_0 – a constant,
T – temperature.

Characteristics $q_{SA}^{T(t)}(T)$ (Fig. 3) shows an inflection point at temperature T_n. By determining tangent to the curve at the inflection point it is possible to establish the activation energy from the following equation [6]:

$$W = -\frac{kT_n^2}{q_{SA}}\frac{dq_S^{T(t)}(T)}{dT}\Big|_{T=T_n} \tag{2}$$

where:

k – Boltzmann constant ($1.38\text{x}10^{-23}$ J/K),
T_n – temperature at inflection point,
$U_n(T_n)$ – voltage at inflection point,
$tg\alpha$ - tangent to the curve at inflection point

For the data obtained from the characteristics shown in Fig. 3 (value of voltage at inflection point U_n = 707 V, temperature at inflection point T_n = 570 K), tangent to the curve at the inflection point and activation energy have been determined as tgα = - 33,33 V/K and W = 1.35 eV, respectively. It has been stated that the lifetime of a charge implanted to dielectric can be assessed based on equation (3), assuming that T_e is a temperature at which the equivalent voltage U_z takes the value of $U_{z\,max}/e$, U_{zmax} is the maximum value of equivalent voltage and e is a base of natural logarithm [6]:

$$\tau(T) \approx \frac{kT_e^2}{bW} \exp \frac{W}{k}\left(\frac{1}{T} - \frac{1}{T_e}\right) \tag{3}$$

where:
T_e – temperature, at which U_z takes the value of $U_{z\,max}/e$,
T –temperature,
b – rate of temperature change during investigation, b = 0,033 K/s.
W – activation energy,
k– Boltzmann constant
The lifetime of the electret charge, based on the data shown in Fig. 3 was evaluated as 10^6 years.

4. Results of measurements

The characteristics of equivalent voltage decay of the charge of electrets made from PTFE foil of 0.5 mm thickness, annealed for four hours at a given temperature and formed either at room T_0 or annealing temperature (Table 1) are shown in Fig. 4. The activation energy and lifetimes τ have been collected in Table 1.

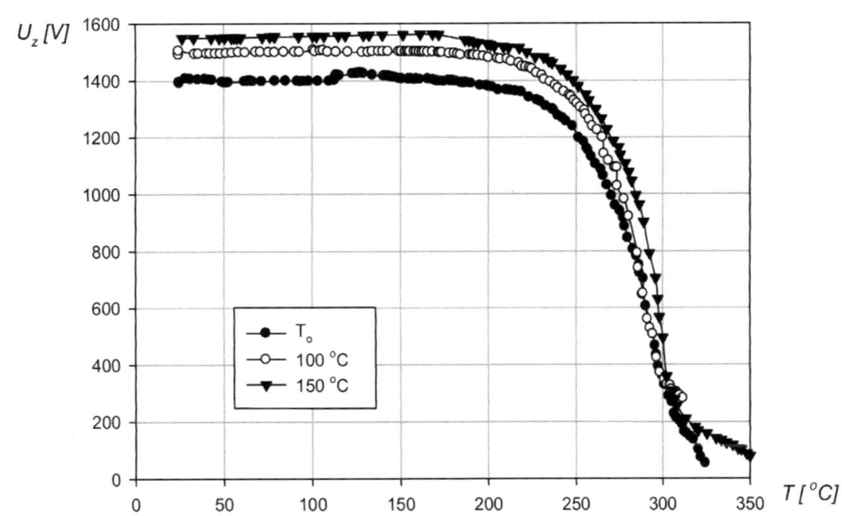

Figure 4. Equivalent voltage versus temperature for different temperatures of electrets forming.

The obtained characteristics provide information about stability of the charge collected in the electret material while the analysis of the decay rate allows us to assess the charge lifetime in the electrets [7].

Table 1. Activation energy and lifetime of electrets

No.	Annealing temperature T_w [°C]	Annealing time [min]	Forming temperature T_f[°C]	T_n [K]	tgα [V/K]	W [eV]	T_e [K]	τ(days)
1.	-	-	T_o	543	8	0,35	563	14,11
2.	100	240	T_o	563	22,2	1,34	568	$2{,}77 * 10^{8}$
3.	100	60	100	533	10	0,34	556	13,05
4.	100	240	100	583	30	1,46	584	$3{,}97 * 10^{9}$
5.	150	60	150	566	10	0,38	566	25,37
6.	150	240	150	587	40	1,47	571	$3{,}59 * 10^{9}$

On the basis of the results collected in Table 1, we can draw the following conclusions:

- Four-hour annealing of the samples at elevated temperature preceding the forming process led to obtaining electrets with a long lifetime (items 2, 4, and 6 in Table 1),
- The value of annealing temperature in the four-hour annealing almost has not changed the resulting lifetime,
- One-hour annealing, independently of the value of temperature, has not extended the lifetime. The lifetime of these electrets is comparable with the lifetime of electrets formed at room temperature, without preliminary annealing.

5. Conclusion

In the paper, results of investigation of selected electrostatic properties of electrets made from PTFE foil of 0.5 mm thickness have been presented. The effects of temperature and annealing time of the samples before the forming process as well as temperature at which charge implantation took place, on the lifetime of the electrets have been analyzed. It has been stated that:

- The lifetime of electrets made from PTFE foil of 0.5 mm thickness depend on the time of samples annealing before their forming,
- The lifetime of the electrets practically does not depend on the temperature of long-term annealing.

6. References

[1] Hilczer B, Małecki 1980 *Elektrety* PWN Warszawa

[2] Kuznetsov A, Rychkov, A, Rychkov D 2011 Electret Properties of Polyethylene and Polytetrafluoroethylene Films with Chemically Modified Surface *IEEE Transactions on Dielectrics and Electrical Insulation* **18** 1

[3] Chen Q 2003 Investigation of corona charge stability mechanisms in polytetrafluoroethylene (PTFE) teflon films after plasma treatment Institute of Plasma Physics, Chinese Academy of Science, People's Republic of China, *Journal of Electrostatics* **59** 3–137

[4] Gang-jin CH, Hui-ming X, Chun-feng Z 2004 Zhejiang University of Science and Technology, *Chen et al. / J Zhejiang Univ SCI* **5** 8 923-927

[5] Łowkis B, Motyl E 1981 Investigations of the organic electrets lifetime, Technical University of Wroclaw *Materials Science* Vol. **VII**, No. 2-3

[6] Van Tournhout J, Ong P H 1977 in: *The investigations of solid dielectrics and methods of their testing* Wrocław 133 - 140

[7] Łowkis B 2010 Efekt elektretowy w dielektrykach polimerowych w aspekcie wykorzystania w medycynie i ochronie zdrowia *Politechnika Wrocławska Monografie* Nr 16

Acknowledgement

The work has been realized within the statutory grant No K0501/S 40029

IOP Conf. Series: Materials Science and Engineering **113** (2016) 012005 doi:10.1088/1757-899X/113/1/012005

The behaviour of water droplets on the silicone rubber surface in an electric field

W Bretuj, A Pelesz
Department of Electrical Engineering Fundamentals, Wroclaw University of Technology. Wybrzeże Wyspiańskiego 27, 50-370 Wrocław, Poland

E-mail: Witold.Bretuj@pwr.edu.pl, Adam.Pelesz@pwr.edu.pl

Abstract. This paper describes the influence of a water droplet placed on flat samples of silicone rubber for enhancement the local electric field and generate electrical discharges. Studies have shown a significant influence of the droplet geometry on the electric strength of the samples. For non-symmetrical arrangement of the three droplets in the inter-electrode space electrohydrodynamic phenomena was observed: a stable change in the droplets shape placed near the electrodes and stretching and tearing down of the water droplets placed far from the electrodes. Captured photos and films of the water droplets behavior placed on the surface of the samples provided data to perform the simulation of the distribution of electric field and an estimate the value of the electric field, which was followed by the development of electric surface discharges.

1. Introduction

Assuring a proper contamination resistance of insulators used in conditions of increased humidity, industrial and environmental pollution is still an important problem not solved entirely. Using composite insulators equipped with housings made of HTV and LSR silicone rubbers in electricity power transmission and distribution lines and covering porcelain insulators already installed with a sheath made of RTV silicone rubber allows substantial decreasing risk of any contamination flashover and increasing reliability of power delivery to consumers.

However, it results from the literature data regarding composite insulators used in power lines that an important danger for these insulators is not an electric flashover, but aging degradation of their surface properties [1,2,3]. Resistance of composite insulators to dangers resulting from their exploitation is connected with their surface properties: strong hydrophobicity of outer housing surface, ability to self-regeneration of surface properties after their temporary loss due to unfavourable environmental conditions and ability to hydrophobization of polluted layer deposited on the insulator surfaces.

In unfavourable weather conditions, long-lasting rains and fogs cause a gradual decline of hydrophobic properties of insulators. Interactions of water molecules with a surface layer of the housing causes turning of methyl groups and approaching the surface by polar linkages of the Si-O chain [4, 5]. This process is reversible. However, temporary moisturizing insulator outer surfaces may initiate surface electric discharges which are a basic factor degrading polymer insulators. Partial electric discharges between wetting areas of the housing deliver energy necessary to reactions creating aging degradation of composite insulators: loss of hydrophobicity, increasing surface roughness and even erosion of the housing material and its unsealing. Exposing a glass - epoxy bearing rod of a composite insulator to exploitation dangers creates danger of breaking the insulator. Problem of aging degradation of the housing material concerns particularly composite insulators of the highest voltage which are characterized by very uneven distribution of voltage along the insulator. In this case, local values of the electric field intensity enhanced by water drops may go much beyond intensity of initiating partial discharges.

The way water wetting housing surfaces behaves influences significantly increase of the local electric field by water drops [6, 7]. Electrohydrodynamic phenomena of deformation, coalescence, instability of a water drop shape and its moving along hydrophobic surfaces of samples in the electric field are a topic of numerous theoretical and experimental works [8, 9, 10].

2. Methodology of research

In this article results of research regarding the influence of how a drop behaves in the electric field on electric endurance of examined samples and distribution of the electric field on their surfaces are presented.

On the long-rod insulator, the direction of the electric field intensity vector is close to tangent at the insulator rod and normal at the shed surface. In the research, a model case of the electric field parallel to the insulating surface was made.

Tests were conducted in the spark gap presented on Figure 1 at constant high voltage of reversed polarity increased with constant speed of about 2 kV/s. The samples were plates made of HTV silicone rubber of 62x40x4 mm stuck centrally between brass electrodes of ø70 mm. An adjustable pipette was used to distribute water drops of defined volume of 80 μl and conductivity of 200 μS/cm on the sample surface.

Figure 1. The measuring spark gap with a stuck sample made of HTV silicone rubber.

Thermal energy and radiation in the area endangered by electric flashover cause temporary degradation of surface properties of silicone rubber. Time for regeneration of surface properties of the samples between succesive tests was 24 hours.

Because of dynamics of observed electrohydrodynamic phenomena, recording of changes in water drop shapes was done using a digital FH100 camera at filming speed of 240 frames per second. Chosen pictures were used to simulate the normalized electric field (for potential 1 V) on the sample surface in the COMSOL MULTIPHYSICS programme.

3. Results of the research

Analysis of behaviour of many water drops located on the solid dielectric surface is very complicated because of mutual corelations between observed electrohydrodynamic phenomena, the electric field distribution and incomplete electric discharges among drops. Taking the above into account, the research was made for chosen variants of distributing three water drops in the interelectrode area.

On Figure 2, chosen pictures of changes in water drop shapes are presented while they are distributed symmetrically in the central part of the interelectrode area. On Figures 2b,c,d,e, are stable changes in shapes of successive water drops situated farther and farther from the high voltage electrode can be seen.

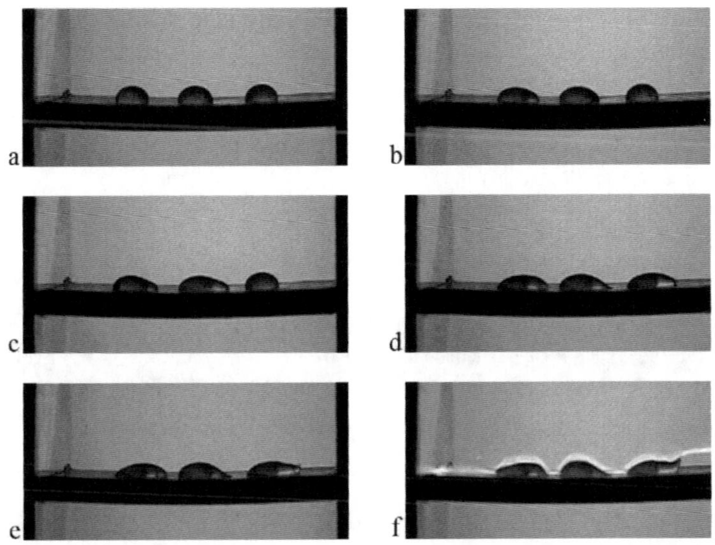

Figure 2. Behaviour of three water drops situated symmetrically in the central part of the interelectrode area: the flashover voltage is 35 kV.

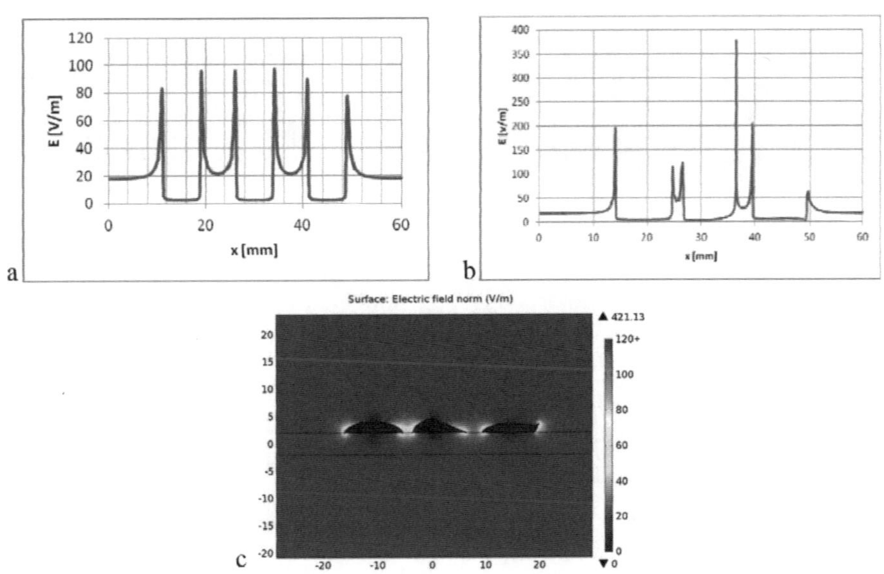

Figure 3. Distribution of the normalized electric field in the tests for water drops situated symmetrically in the central part of the interelectrode area: a) droplets not deformed, b, c) a moment before the flashover

Distribution simulations of the normalized field made for drops which are not deformed in a electric field (Fig. 3a) and a moment before the flashover (Fig. 3b) indicate more than threefold increase of the local electric field values caused by stretching and shape changes of the drops. At relatively small drop conductivity of 200μS/cm, the way of flashover goes through the drop surface (Fig. 2f) and joins the strong electric field areas as shown on Figure 3c.

For Figure 4a presenting configuration of water drop distribution during increase of the test voltage, initially strong stretching of the water drop situated in the middle of interelectrode area in the direction of the earthed electrode was captured (Fig. 3b). Such a drop deformation caused almost twofold increase of the electric field value at its rim (Fig. 5). While the test voltage was increasing further, the drop came back to the original shape (Fig. 4d) and oscillated in the electric field (Fig. 4f, g). The cause of vibrations could be the electric charge of the middle drop because of partial discharges to the drops at the electrodes.

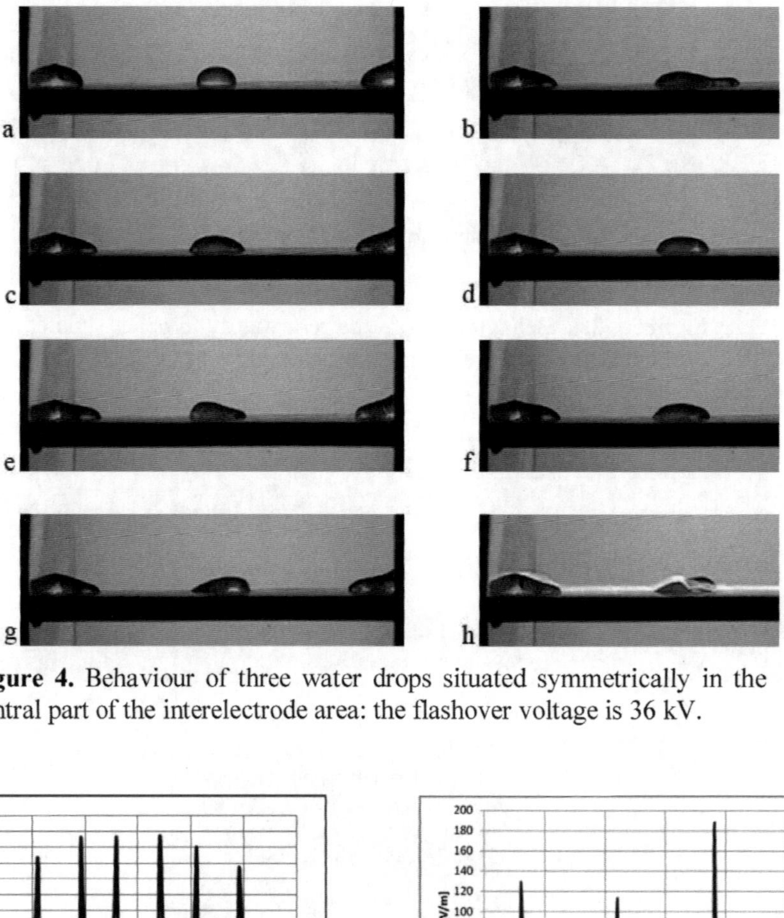

Figure 4. Behaviour of three water drops situated symmetrically in the central part of the interelectrode area: the flashover voltage is 36 kV.

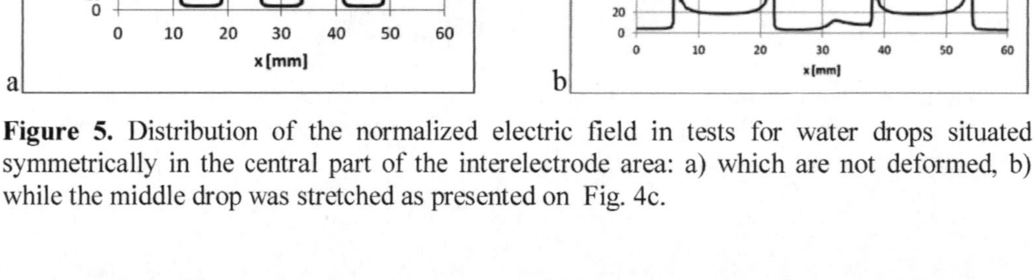

Figure 5. Distribution of the normalized electric field in tests for water drops situated symmetrically in the central part of the interelectrode area: a) which are not deformed, b) while the middle drop was stretched as presented on Fig. 4c.

IOP Conf. Series: Materials Science and Engineering **113** (2016) 012005 doi:10.1088/1757-899X/113/1/012005

In the research made for a system of three drops situated not symmetrically in the interelectrode area, the electric sample endurance was dependent on polarity of the electrode near which the drops were situated. Figure 6 presented electrostatic distribution of the normalized electric field for water drops not deformed in the electric field dependent on geometrical configuration.

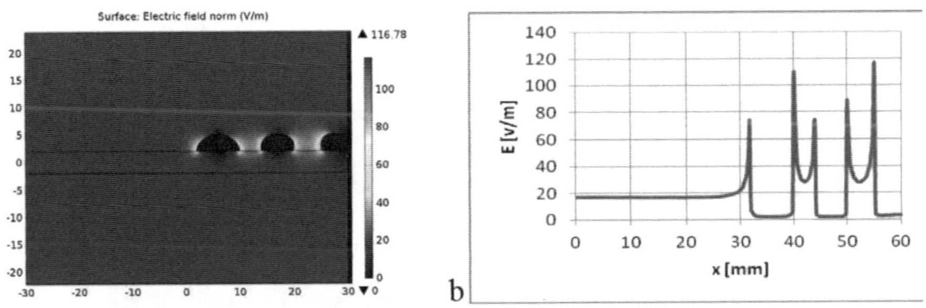

Figure 6. Distribution of the normalized electric field in tests for water drops situated in the half of the interelectrode area; the drops are not deformed in the electric field.

The flashover voltage value measured at the constant voltage for drop distribution near the high voltage electrode was 26 kV. Figure 7 presents chosen recordings of drop behaviours while increasing test voltage values and the electric flashover between the electrodes at the voltage of 26 kV (Fig. 7f).

Stretching of drops was observed before the flashover. The drop moved most to the interelectrode area was deformed most. Its sudden stretching in the direction of the earthed electrode was recorded before the flashover.

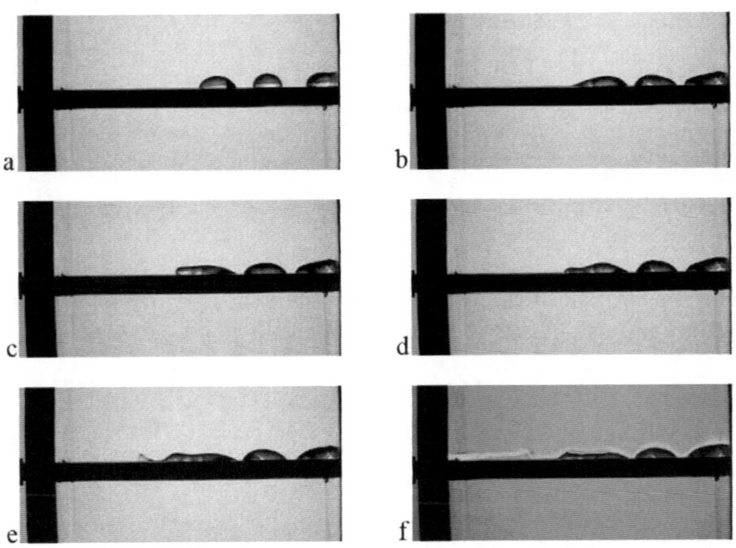

Figure 7. Behaviour of three water drops situated near the high voltage electrode: the flashover voltage is 26 kV.

Figure 8 presents electrohydrodynamic phenomena recorder for drops situated near the earthed electrode. The electric flashover happened at the voltage of 22 kV. Stable stretching of the water drops was observed before the flashover.

Charts presenting distribution of the normalized electric field for samples with water drops situated in the half of the interelectrode area are shown on Figure 9. Maximum increase of the local electric field value caused by drop deformation before the flashover did not depend on the electrode polarity at which water drops were situated and was similar to values shown on Figure 6b for drops which were not deformed.

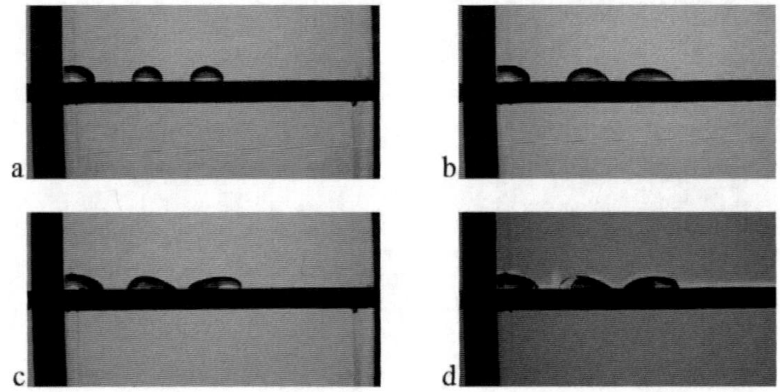

Figure 8. Behaviour of water drops situated near the earthed electrode: the flashover voltage is 22 kV

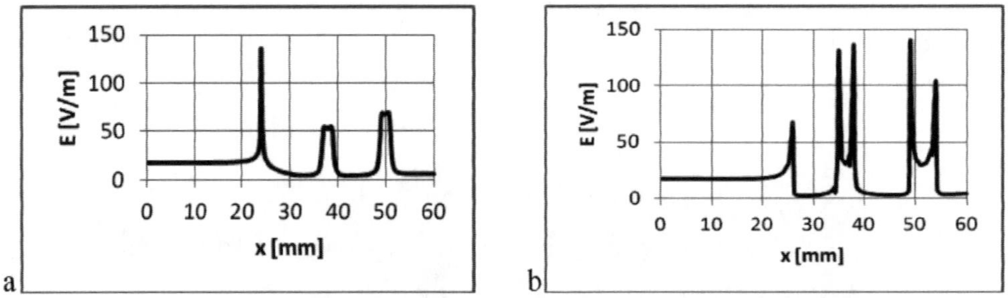

Figure 9. Distribution of the normalized electric field before the flashover for samples with water drops situated in the half of the interelectrode area: a) at the high voltage electrode, b) at the earthed electrode

Electric endurance of samples with water drops situated near the high voltage electrode was approximately 15% lower than samples with water drops situated near the earthed electrode.

4. Summary

At symmetrical distribution of water drops on the sample surfaces, deformations of water drops shapes have a strong influence on distribution of the electric field on sample surfaces. Maximum values of the local electric field at water rims before the flashover are two- or threefold higher than for drops which are not deformed. There is no important influence observed of contact of drops with the electrodes on increasing the electric field and the flashover voltage on the samples.

Concentration of water drops near one of electrodes influenced decrease of the sample electric endurance by about 30% in comparison with samples moisturized symmetrically.

At asymmetrical distribution of water drops on the sample surfaces, observed electrohydrodynamic phenomena cause a small increase of the maximum electric field intensity value by about 20%. Polarity of the electrode at which drops were situated has influence on the flashover voltage on the samples. In the research made, electric endurance of samples with water drops situated near the high voltage electrode (DC-) was approximately 15% lower than the samples with water drops situated near the earthed electrode.

The research proved that geometrical configuration of water drop distribution on the hydrophobic sample surfaces in the constant electric field has a substantial influence on electrohydrodynamic phenomena, distribution of the electric field and electric endurance of the samples.

5. References

[1] Gorur R S, Karady G G, Jagota A, Shah M, Yates A M 1992 Aging in silicone rubber used for outdoor insulation *IEEE Transactions on Power Delivery* **7** 2 525-531

[2] Goudie J L, Owen MJ, Orbeck T 1998 A review of possible degradation mechanisms of silicone elastomers in voltage insulation applications *Conference on Electrical Insulation and Dielectric Phenomena Annual Report.* **1** 120 – 127

[3] Reynders J P, Jandrell I R, Reynders S M 1999 Surface ageing mechanisms and their relationship to service performance of silicone rubber insulation *High Voltage Engineering Symposium* 22-27 August 1999

[4] Goudie J L, Owen M J, Orbeck T 1998 A review of possible degradation mechanisms of silicone elastomers in voltage insulation applications *Conference on Electrical Insulation and Dielectric Phenomena, Annual Report.* **1** 120 – 127

[5] Yoshimura N, Kumagai S 1999 Electrical and environmental aging of silicone rubber used in outdoor insulation *IEEE Transactions on Dielectrics and Electrical Insulation* **6** 5 632-650

[6] Zhicheng G, Liming W, Bo Y, Xidong L, Zhi L 2005 Electric Field Analysis of Water Drop Corona *IEEE Transactions On Power Delivery* **20** 2 964 – 969

[7] Marungsri B, Onchantuek W, Oonsivilai A, Kulworawanichpong 2009 Analysis of Electric Field and Potential Distributions along Surface of Silicone Rubber Insulators under Various Contamination Conditions Using Finite Element Method *World Academy of Science Engineering and Technology* 53

[8] El-Kishky H, Gorur R S 1996 Electric Field Computation on an Insulating Surface with Discrete Water Droplets *IEEE Transactions on Dielectrics and Electrical Insulation* **3** 3 450-456

[9] Imano A M, Essiane S N, Beroual A 2007 Simulation of the shapes of a water droplet on insulated solid surface in an AC electric field *Published on the NRC Research Press Web site at http://cjp.nrc.ca/ on 22 August 2007*

[10] Jiang X, Du B X, Liu Y, Zhu X 2005 Dynamic behavior of water droplet on polymer insulator under DC field *Proceedings of the XIVth International Symposium on High Voltage Engineering, Tsinghua University Bejing, Chin 2005*

Acknowledgement

This publication was prepared within the statutory research by Department of Electrical Engineering Fundamentals, Wroclaw University of Technology

IOP Conf. Series: Materials Science and Engineering **113** (2016) 012006 doi:10.1088/1757-899X/113/1/012006

Microplasma technology for carbon micro- and nanostructures formation in gas at atmospheric pressure

A T Sobczyk, A Jaworek, M Klein
Institute of Fluid Flow Machinery, Polish Academy of Sciences, Fiszera 14, Gdańsk, POLAND

sobczyk@imp.gda.pl

Abstract. The paper shows the results of investigation of various carbon microstructures deposited on electrodes in low-current electrical-discharges in cyclohexane as carbon feedstock and argon as neutral gas at atmospheric pressure. The plasma was generated between stainless steel needle and plate made of nickel alloy. The carbon micro- and nanostructures have grown at the tip of the needle electrode (anode) for the discharge current ranged from 0.1 mA to 3 mA. The physical properties of carbon fibres were investigated by SEM, EDS and Raman spectroscopy.

1. Introduction

Electrical discharge as a source of plasma is applied in various industrial processes, for example, for surface modification, sterilization, noxious compounds decomposition, particulate matter precipitation, micro- and nanostructures synthesis, plasma etching, or electric discharge machining. These processes, operating at pressures ranging from atmospheric to below 1 Pa, are conducted between electrodes of different geometries, by DC or AC high voltage of various frequencies, or by microwave excitation.

The term *microplasma* refers to a discharge, in which at least one of linear dimensions of the plasma active region is of tens to a couple of thousands microns. Usually microplasma is generated at gas of near atmospheric pressure (0.1 MPa). Main advantage of microplasma-based processes, compared to low-pressure plasma sources, is that chemical reactions can proceed at high-pressure gas flowing continuously through plasma reactor of small size. In materials processing science, microplasma is used to synthesize nanoscale structures such as, for example, carbon dendrites [1], SiO_2 thin films [2], Si nanocones, WOx nanoparticles, diamond crystals [3], or carbon amorphous deposits [4]. Generally, the most popular discharge types for microplasma generation are dielectric barrier discharge and plasma-jet discharge. The aim of the present paper is to present experimental results of investigation of microplasma generated by corona discharge in needle-plate configuration used for the synthesis of various carbon structures.

2. Experimental

A schematic of experimental set-up is shown in Figure 1. The experiments were carried out in a reactor chamber of 0.1 dm^3, made of PMMA, at normal pressure, in the atmosphere of argon as carrier gas and cyclohexane as carbon feedstock. The hydrocarbon vapours were delivered to the reactor using a bubbler, in which gaseous bubbles were produced by argon flowing through it. The temperature of liquid cyclohexane was stabilized at 25° C. The concentration of hydrocarbon in the gas mixture was controlled by mixing the vapours leaving the bubbler with additional argon. Both

IOP Conf. Series: Materials Science and Engineering **113** (2016) 012006 doi:10.1088/1757-899X/113/1/012006

flow rates of argon were controlled by two rotameters. The concentration of C_6H_{12} in argon, determined from the saturation vapour pressure in the bubbler and the flow rate of additional argon, was estimated to 3.5 at%. The flow rate of the gas mixture was 0.05 dm^3/min.
The discharge was generated between a stainless steel needle and a plate made of nickel alloy. The diameter of the needle was 1 mm, and the dimensions of the plate were 35x20 mm. The distance between the electrodes was 15 mm.
The discharge electrodes were supplied from high voltage DC source SPELLMAN HV SL 600W/40kV/PN of positive polarity. The discharge current was stabilized by a series resistor of the value of 5 MΩ. The discharge current was changed from 0.1 mA up to 3 mA. The time of plasma reactions in the reactor in these experiments was usually 30 s, but for low discharge current of 1.4 mA, the time was 5 minutes.
Gaseous products leaving plasma reactor were analysed using gas chromatography method. The as-made carbon deposit grown on the needle tip was carefully removed from the reactor and examined under scanning electron microscope (Zeiss EVO 40) equipped with energy dispersive spectroscope (EDS, Bruker). Next, Renishaw inVia Raman Microscope with a 100x objective lens, with specimen excitation by laser beam of wavelength of 514 nm and of power of 1.5 mW was used in order to analyse the Raman spectra.

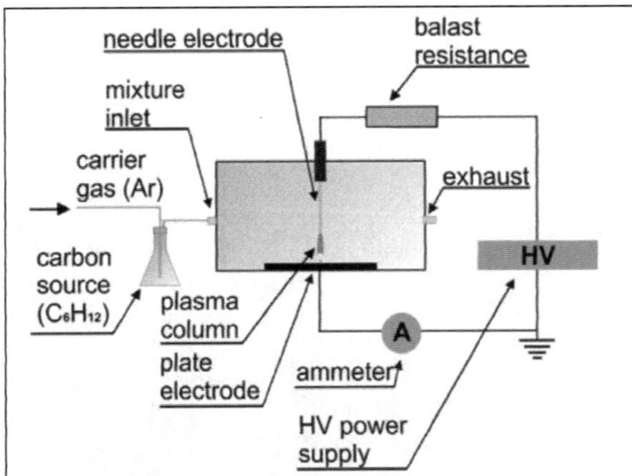

Figure 1. Experimental set-up for micro- and nanostructures synthesis in Ar:C_6H_{12}.

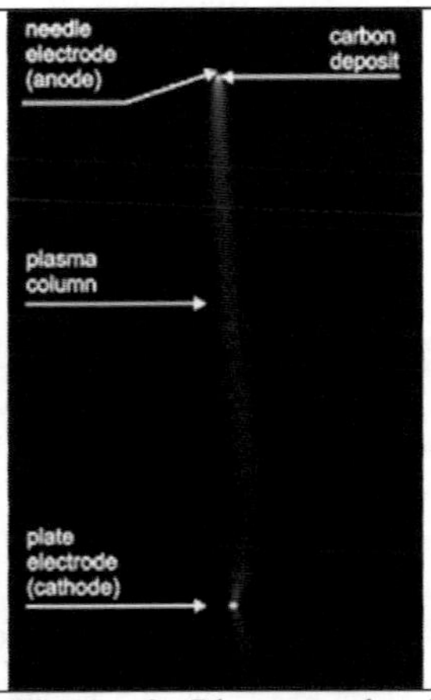

Figure 2. Photograph of plasma column of low-current electrical discharge, in which carbon deposits are synthesized (exposure 10 ms).

3. Results

Synthesis of carbon micro- and nanostructures was carried out in low current, high voltages electrical discharge in Ar:C_6H_{12} mixture at atmospheric pressure. The type of discharge generated under these conditions was glow discharge with visible plasma column in the form of thin filament bridging the anode and the cathode. Figure 2 presents a photograph of plasma column for a discharge current of 1.8 mA.
The power of discharge for currents in the range from 1.4 mA to 1.9 mA was about 5 W and did not change significantly with the current increasing (Figure 3). For discharge currents higher than 1.9 mA, the discharge power increased almost linearly from 6 to about 10 W. The maximum value of specific input energy, defined as the ratio of average discharge power to the flow rate of gaseous molecules in the plasma volume, was about 95 eV/molecule, for a discharge current i=1.4 mA (Figure 4). The minimal value of specific input energy, was obtained for the discharge current of 1.8 mA.
Main gaseous products obtained during the discharge in mixture of Ar and c-C_6H_{12} were ethylene, methane and acetylene. The concentration of other neutral gaseous products was below 0.1 percent.

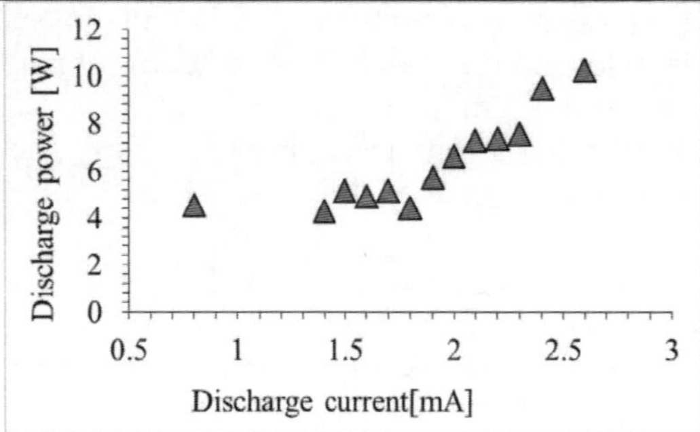

Figure 3. Dependence of discharge power on discharge current.

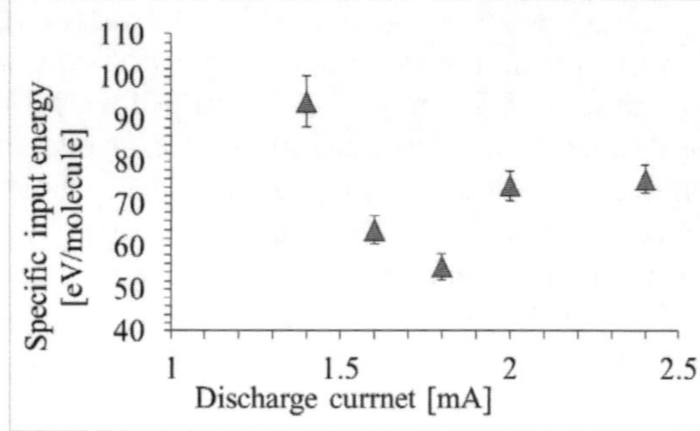

Figure 4. Dependence of specific input energy on discharge current.

The morphology of deposits obtained on the needle electrode was different depending on the discharge current (Figure 5). For a current of 0.6 mA, the carbon deposit was of irregular structure (Figure 5a). For a current of 1.4 mA, the deposit was in the form of filament covered with carbon nanowalls (Figure 5b). It was observed that the nanowalls deposit was formed only on the side-wall of the filament, far from its tip. For discharge currents in the range from 1.7 to 2.0 mA, the deposit had a shape of regular filament with diameter in the range from about 10 to 30 μm (Figure 5c). The surface of this filament was smooth, and its cross section was homogeneous. For discharge currents higher than 2.0 mA, the surface of filament was covered with irregular pebble-like soot deposit.

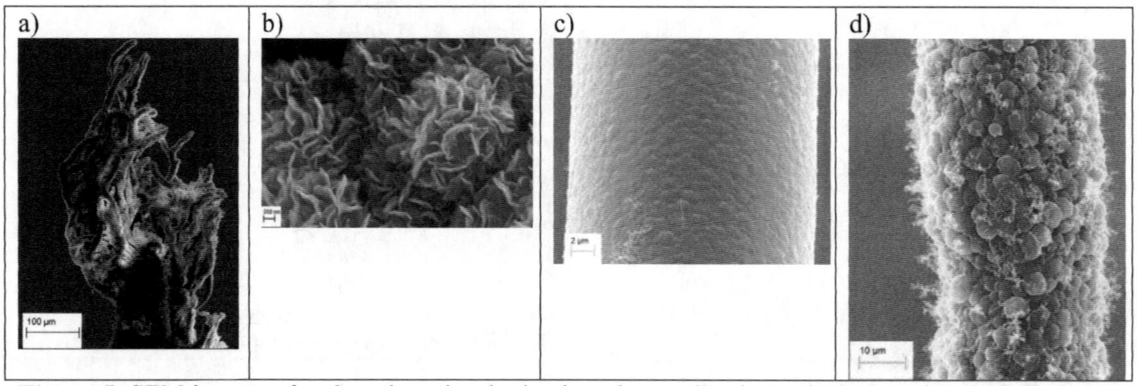

Figure 5. SEM images of carbon deposits obtained on the needle electrode during electrical discharge in a mixture of argon and cyclohexane for various discharge currents a) 0.1 mA, b) 1.4 mA, c) 1.8 mA, d) 2.4 mA.

The elemental analysis confirmed that carbon is the dominant element in the produced filament. Comparing the concentration of carbon and hydrogen in cyclohexane (C — 85.7 wt.%, H — 14.3 wt.%) with those in the deposits, it is evident that highly carbonised products (C — 98.7 wt.%, H — 0.9 wt.%, the rest was N and O absorbed from the atmosphere) can be obtained from the decomposition of cyclohexane in electrically generated microplasma.

For all discharge currents, except that of 0.1 mA, the plate electrode (cathode) was covered by soot and graphitized small carbon choke. For this current (0.1 mA), the plate was also covered with "oily" substance, which was probably higher hydrocarbons synthesised during electrical discharge.

The Raman spectra obtained for the deposit produced by a current of 1.8 mA (Figure 6) are similar to those measured for a commercial glassy carbon (supplied from Alfa Aesar). The G band and D band positions are similar and are located at 1592 cm^{-1} and at 1347 cm^{-1}, respectively. The difference in graphitization degree between glassy carbon and deposit obtained during the discharge can be estimated from the ratio of intensity of the G and D bands. The results show that the deposit obtained in these experiments has a relatively higher graphitization degree compared to the commercial glassy carbon, because this ratio was higher. There was no significant difference in Raman spectra between deposits obtained for discharge current 1.4 mA and 1.8mA.

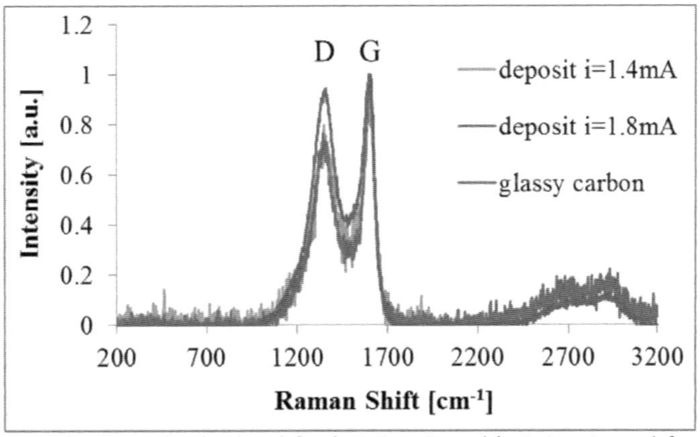

Figure 6. Raman spectra for deposit obtained for i=1.8 mA and i=1.4 mA, and for glassy carbon.

4. Discussion

The main building blocks of carbon deposits obtained in the discharge are radicals and neutral products of decomposition of cyclohexane. These products are obtained due to collision with electrons of sufficiently high energy, for example:

$$c - C_6H_{12} + e \rightarrow R_m^{\bullet} + R_n^{\bullet} + e$$

where R_m and R_n are radicals.

Similar situation occurs when argon atom in the first excited state (metastable state) transfers its internal energy to cyclohexane molecule after impact:

$$c - C_6H_{12} + Ar_m \rightarrow Ar(^1S_0) + R_m^{\bullet} + R_n^{\bullet}.$$

Because the energies of the argon metastable states are relatively high, 11.55 eV and 11.72 eV, the cyclohexane and all products of cyclohexane decomposition can be ionized or dissociated to other products. Radicals could interact with each other forming neutral molecules and/or other new radicals. Production of ions by electron impact is also possible. The ionization energy of cyclohexane is about 9.9 eV. Very close to this value are also energies of some dissociation processes in cyclohexane by which such ions as $C_5H_9^+$, $C_4H_8^+$, $C_4H_7^+$, $C_6H_{11}^+$ and $C_3H_7^+$ are formed [5-7]. Jiao and Adams [8] measured the production rate of ions due to collision of electrons with cyclohexane molecules, and discovered that the most popular ion after $C_6H_{12}^+$ is $C_4H_8^+$, and the third one is $C_5H_9^+$. During the synthesis of carbon structures in electrical discharge in a mixture of cyclohexane and argon, the ethylene had the highest concentration from all gaseous products.

Positive hydrocarbon ions produced in the discharge cannot reach anode to build a deposit on it, but can become target for electrons. Due to radiative or dissociative electron attachment, these ions can become a source of lighter hydrocarbons, radicals or even negative ions, for example, CH_2^- or C_2H^-.

The formation of "oil" form of deposit remaining on the plate electrode can proceed due to the following reaction:

$$R_x^{\bullet} + R_y^{\bullet} \rightarrow P_{x+y}$$

where P_{x+y} is a neutral product (hydrocarbon) of higher molecular weight than cyclohexane.

Carbon nanowalls were obtained for conditions for which the specific input energy of about 90 eV/molecule was close or higher than the sum of all bond energies in cyclohexane (about 74 eV). These conditions allowed obtaining C, C_2 and CH_3 molecules which are known as building blocks of carbon structures. Carbon nanowalls, which are highly graphitized form of carbon, were formed on the sidewall of growing carbon fiber.

Deposit obtained for 1.8 mA was probably built similarly to other amorphous forms of carbon (diamond-like carbon) , i.e., during CH_3 layer-by layer deposition onto the needle. Conditions obtained for discharge current higher than 2.0 mA (resulting in higher methane and acetylene concentrations) favoured nucleation of soot precursor and carbonaceous powder in the space between electrodes. Particles formed in this discharge usually had negative charge

$$C_2H_2 + e^- \rightarrow C_2H^- + H$$

$$C_{2n}H_2 + C_2H^- \rightarrow C_{2n+2}H^- + H_2, n = 1,2,3 \ldots$$

This process is known as "Wichester mechanism" [9]. Negative ions tend to flow towards carbon deposit located on positively polarized needle.

4. Conclusions

In microplasma generated by electrical discharge in a mixture of argon and cyclohexane, different forms of carbon structures were obtained. It was shown how different forms of carbon deposit depend on the magnitude of discharge current, which ranged from 0.1 mA to 3 mA in these specific experimental conditions. The carbon structures were effectively synthesised in discharge of positive polarity generated between a stainless steel needle and a plate made of nickel alloy by 15 mm interelectrode distance. Depending on the discharge current, carbon nanowalls, glassy carbon or smooth carbon filaments, which can be useful for various applications, were produced. Carbon nanowalls, for example, are potentially applicable in fuel cells, chemical sensors, biological sensors and biomedical devices.

Acknowledgements. The work was supported by the Institute of Fluid Flow Machinery, Polish Academy of Sciences within the project No. O1/T3/Z4. The support of Anna Białous of the Institute of Fluid Flow Machinery in spectral measurements is gratefully acknowledged. The authors wish to thank Mr. Jerzy Bunikowski for his technical help in construction of experimental stand.

References

[1] Kozak D., Shibata E., Iizuka A., Nakamura T., Growth of carbon dendrites on cathode above liquid ethanol using surface plasma, Carbon 70 (2014) 87 –94

[2] Gherardi N., Martin S., Massines F., A new approach to SiO_2 deposit using a N_2 –SiH_4 –N_2O glow dielectric barrier-controlled discharge at atmospheric pressure, J. Phys. D: Appl. Phys. 33 (2000) L104–L108

[3] Shirai H., Kobayashi T., Haegawa Y., Synthesis of silicon nanocones using rf microplasma at atmospheric pressure, Applied Physics Letters 87 (2005), 143112

[4] Kikuchi T., Hosegawa Y., Shirai H., RF microplasma jet at atmospheric pressure: characterization and application to thin film processing, J. Phys. D: Appl. Phys. 37 (2004) 1537–1543

[5] Lossing, F.P.; Traeger, J.C., Free radicals by mass spectrometry XLVI. Heats of formation of C_5H_7 and C_5H_9 radicals and cations., J. Am. Chem. Soc., 1975, 19, 9

[6] Rabbih, M.A.; Selim, E.T.M.; Fahmey, M.A., Ionization & fragmentation of cyclohexane, Indian J. Pure Appl. Phys., 1981, 19, 962

[7] Sergeev, Yu.L.; Akopyan, M.E.; Vilesov, F.I.; Chizhov, Yu.V., Photoionization processes in gaseous cyclohexane, and chloro- and bromocyclohexane, High Energy Chem., 1973, 7, 369, 418

[8] Jiao C.Q., Adams S.,F., Electron ionization of selected cyclohexanes, J. Phys. B: At. Mol. Opt. Phys. 44 (2011) 175209

[9] De Bleecker K., Bogaerts A., Goedheer W., Modelling of nanoparticle coagulation and transport dynamics in dusty silane discharges, New J. Phys. 8 (2006), 178.

Thin-film $TiPbO_3$ varistors obtained by two-source magnetron sputtering

J Ziaja, M Lewandowski
Department of Electrical Engineering Fundamentals,
Wroclaw University of Technology, Poland

E-mail: jan.ziaja@pwr.edu.pl

Abstract. The paper presents the method of obtaining thin films of $TiPbO_3$ by two-source magnetron sputtering DC-M. The films were obtained in a reactive process of sputtering metallic targets of titanium (Ti) and lead (Pb). The research involved the impact of the time of sputtering of the respective targets on voltage-dependent resistance of the obtained films for different power conditions, pressures of process gases and the powers provided on the targets. The obtained nonlinearity coefficients and the current-voltage *I*(*U*) characteristics were within the following range.

1. Introduction

These Varistors are electrical elements characterised by extreme nonlinearity of the current-voltage (*I*-*U*) dependency. As such, they are widely used as protection devices against surges in electrical and power equipment. The varistors that are manufactured on an industrial scale include a film of sintered zinc and other metal oxides with a thickness between several millimetres to several centimetres. The power energy industry uses varistors with a wide range of actuation voltage – from several volts into thousands. The manufacture of varistors with increasing reliability and uniform microstructure and resistivity distribution is still a challenge for researchers. A number of sophisticated methods have been employed to obtain such varistors: advancing technological processes, applying chemical methods of powder uniformisation, or introducing optimalisation formulas based on network modelling techniques. The available varistors can protect systems and circuits in a wide range of voltages – from several volts in case of low-voltage varistors present in electronic circuits, to several thousand volts in power networks. The manufacture of varistors with increasing reliability and uniform microstructure and resistivity distribution is still a challenge for researchers.

Non-linear electric conductance in ceramic varistors is made possible by the barriers to electric potential at the perimeters of the granules, the presence of which is related to the energetic and spatial distribution of trapping states on interfacial surfaces [1, 2]. The form of the current-voltage characteristics *I*(*U*) points to the existence of a several underlying mechanisms related to electrical conductivity, and is a function of both the chemical composition and the heat treatment conditions [3]. A number of sophisticated methods have been employed so as to obtain such varistors: by advancing technological processes, by applying chemical methods of powder uniformisation, or by introducing optimalisation formulas via network modelling techniques [4]. The available varistors can protect systems and circuits in a wide range of voltages – from several volts in case of low-voltage varistors present in electronic circuits, to several thousand volts in power networks. Today, research efforts focus on developing new materials and methods of manufacture of voltage-dependent resistance materials. Lead titanate ($TiPbO_3$) is one of such promising materials. It shows ferroelectric and piezoelectric properties, and voltage-dependent resistance. Magnetron sputtering is the most beneficial method of obtaining thin-film structures. Magnetron systems make it possible to clean and harden the substrate, activate the surface, and to apply films within a single process. The films obtained in this manner are characterised by decent hardness and adhesion to the substrate in relation to the original material.

2. Experiment

Thin films of $TiPbO_3$ were obtained in the process of reactive sputtering of metallic targets of lead and titanium with a purity of 99.99%. Two WMK-50 magnetron launchers were used (Fig. 1). The partial pressure of oxygen and argon were identical and amounted to $p_{O2} = p_{Ar} = 1.33$ Pa, thus the final pressure of the process mixture amounted to p = 2.66 Pa. The magnetron launchers are powered with a DPS (DORA POWER SYSTEM) pulsed current source with a maximum output of 16 kW and a maximum voltage of 1.2 kV. The group frequency of the source was adjustable in the range from 5 Hz to 5 kHz. The use of a pulsed current source (Fig. 2) makes it possible to vary the load concentration in the area between the target and the substrate, since cyclical changes in the electric force rearranges ions/electrons in the area of the influence of the force. During the current pulse time t_1 , process gas ions actively sputter the target, whereas the electric and magnetic field holds it within the plasma area. The disappearance of the electric field (between the pulses t_2-t_1) releases ions/atoms that condense on the substrate thus forming the film.

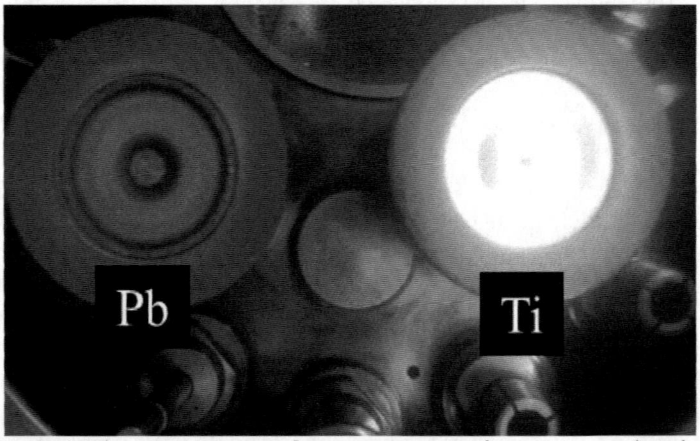

Figure 1. The arrangement of two magnetrons in a vacuum chamber.

Due to the different points of melting (lead: 327 °C, titanium: 1688 °C), boiling (lead: 1749 °C, titanium: 3287 °C) and different heat conductivity (Ti-22,4 $W m^{-1} K^{-1}$, Pb-35,3 $W m^{-1} K^{-1}$) it is difficult to sputter these elements using the same power densities and over the same periods of time. To overcome this difficulty a muliplexer was designed that powers both magnetron launchers in an alternating manner. The multiplexer makes it possible to vary the switching frequency and sputtering time depending on the target (Fig. 3). This solution makes it possible to obtain single layers of $TiPbO_3$, as well as structures made of alternating layers of TiO_2 and PbO. This, in turn, makes it possible to influence the size of connections and interfaces between the granules. The switching frequency of the targets was set at 5 Hz for the purposes of the research presented in this paper. The total time of applying layers amounted to 30 min. Maximum power provided on the targets amounted to 300 W. The distribution of sputtering time between the respective targets ranged varied from 100% to 25%. The distance between the target and the substrate amounted to 10 cm.

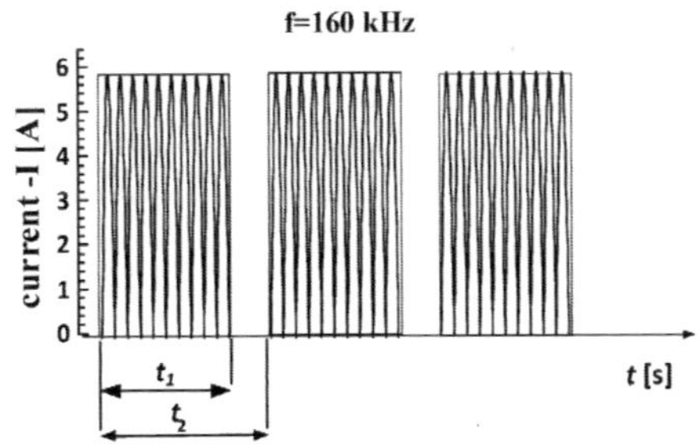

Figure 2. Drawings of rectangular modulated discharge current pulse d.c. mode bias DC-M.

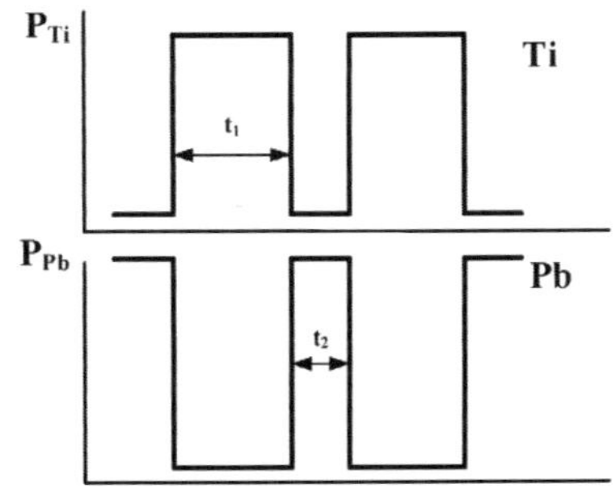

Figure 3. Time of sputtering the respective targets: t_1 – Ti target sputtering time, t_2 – Pb target sputtering time.

3. Results

The current-voltage characteristics were measured with a Keithley 617 electrometer with a voltage increase of 100 mV/min. Fig. 4 presents a comparison of the measured current-voltage characteristics of the obtained voltage-dependent resistance films. It can be observed that the TiO_2 layer is characterised by strong nonlinearity of $I(U)$. The calculated nonlinearity coefficient amounted to α=24. On the other hand, in case of PbO layers this relation is nearly linear. Nonlinearity occurs for polarisation voltage above 1 V. Figure 5 shows the $I(U)$ characteristics of TiO_2 layers doped with PbO. The doping was effected by changing the distribution of the sputtering time of titanium and lead metallic targets. Lead and titanium targets were sputtered with a power of 300 W and with a switching frequency of 5 Hz. It can be noted that doping substantially lowers the nonlinearity coefficient - down to the 4 to 10 range.

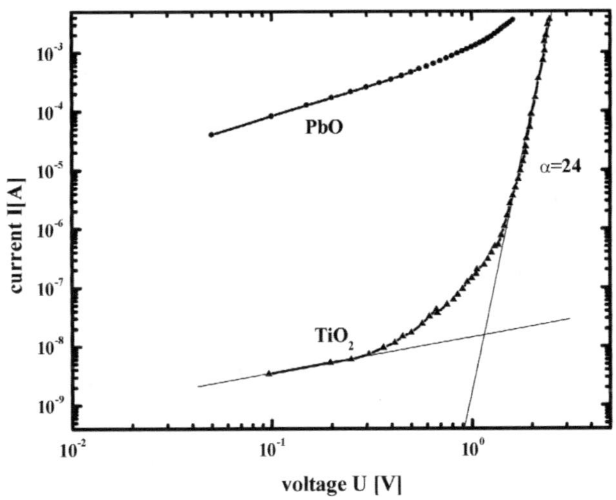

Figure 4. Current as a function of voltage for PbO and TiO_2 layers.

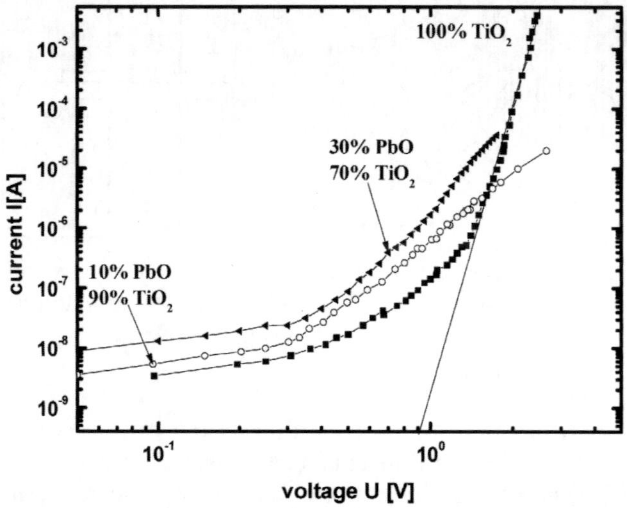

Figure 5. Current as a function of voltage for $TiPbO_3$ layers at various power provided at targets

4. Summary

The TiO_2 layers obtained by magnetron sputtering show voltage-dependent resistance with high nonlinearity coefficient α=24. The layers of PbO do not show voltage-dependent resistance. By adjusting the time and power of sputtering the targets, and the switching frequency it is possible to increase the nonlinearity coefficient. The actuation voltage of the voltage-dependent resistance layers can be adjusted by varying their thickness (by regulating the total sputtering time).

References

[1] Amrani B and Hamzaoui S 2004 *Characterization of ZnO films prepared by reactive sputtering at different oxygen pressures sputtering at different oxygen pressures*, Catalysis Today 89, p 331–335.

[2] Artaud M C, Ouchen F, Martin L and Duchemin S 1998 *CuInSe$_2$ thin films grown by MOCVD: characterization, first devices*, Thin Solid Films 324 p 115–123

[3] Liu C Y, Zhang B P, Binh N T and Segawa Y 2004 *Third-harmonic generation from ZnO films deposited by MOCVD*, Applied Physics B, Lasers and Optics p 83–86

[4] Miśta W, Ziaja J and Gubański A 2004 *Varistor performance of nanocrystalline Zn–Bi–O thin films prepared by reactive RF magnetron sputtering at room temperature,* Vacuum, , Vol. 74, p 293–296.

Research carried out within the statutory activities of the Department of Electrical Engineering Fundamentals

IOP Conf. Series: Materials Science and Engineering **113** (2016) 012008 doi:10.1088/1757-899X/113/1/012008

Equivalent circuit of a coaxial-line-based nozzleless microwave 915 MHz plasma source

R Miotk[1], M Jasiński[1] and J Mizeraczyk[2]
[1]*Centre for Plasma and Laser Engineering, The Szewalski Institute of Fluid-Flow Machinery Polish Academy of Sciences, ul. Fiszera 14, 80-231 Gdańsk, Poland*
[2]*Department of Marine Electronics, Gdynia Maritime University, ul. Morska 81-87, 81-225 Gdynia, Poland*

E-mail: rmiotk@imp.gda.pl

Abstract. This paper presents a new concept of an equivalent circuit of a microwave plasma source (MPS) used for gas treatment. The novelty of presented investigations is the use of the Weissfloch circuit as equivalent of an area of waveguide discontinuity in the MPS which is a result of entering a coaxial-line structure. Furthermore, in this area the microwave discharge is generated. Verification of the proposed method was carried out. The proposed equivalent circuit enabled calculating the MPS tuning characteristics and comparing them with those measured experimentally. This process allowed us to determine the impedance Z_P ofplasma in the MPS.

1. Introduction

To meet industrial requirements of gas treatment presented MPS has to be optimized in terms of efficiency of the power transfer from the feeding microwaves to the plasma. This efficiency is usually expressed as the ratio P_R/P_I, where P_I and P_R are the powers of the incident and reflected microwaves at the input of the MPS, respectively [1]. The tuning characteristics of the MPS is the dependence of P_R/P_I on the normalized position of the movable plunger l_S/λ_g, i.e. P_R/P_I (l_S/λ_g), where l_S is distance between the MPS output plane and the movable plunger (Figure 1) and λ_g is the wavelength of microwaves in the waveguide (in our case, at frequency 915 MHz and waveguide type WR 975 the λ_g is equal to 437.7 mm). The MPS operation is efficient when the value of ratio P_R/P_I is low and does not depend strongly on the movable plunger position [1, 2].

In this paper we present the equivalent circuit of the waveguide-supplied coaxial-line-based nozzleless MPS used for gas treatment at high flow rate [1]. The equivalent electric circuit concept serves to analyse the characteristics of the MPS without necessity of solving the full set of electromagnetic field equations [2]. The novelty of the presented equivalent circuit lies in using the Weissfloch circuit [3]. The proposed equivalent circuit of the MPS allowed to determine the impedance Z_P of the plasma and calculate the dimensions of a diaphragm which increases the efficiency of microwave power transfer from the feeding microwaves to the plasma.

2. The coaxial-line-based nozzleless MPS

A sketch of the MPS is shown in Figure 1. The MPS was described in details in [1]. The MPS, operating at the atmospheric pressure, was based on the standard WR 975 rectangular waveguide with a section of reduced-height. Before this section a metal tapered section was placed. The coaxial-line structure (consisted of coaxial quartz, inner electrode and metal tube) extends through the wider wall of the reduced-height waveguide section (plane 3-3'). The microwave power was supplied to the MPS input plane (1-1') also via the standard rectangular waveguide WR 975. At the MPS output plane 4-4' a movable plunger was attached.

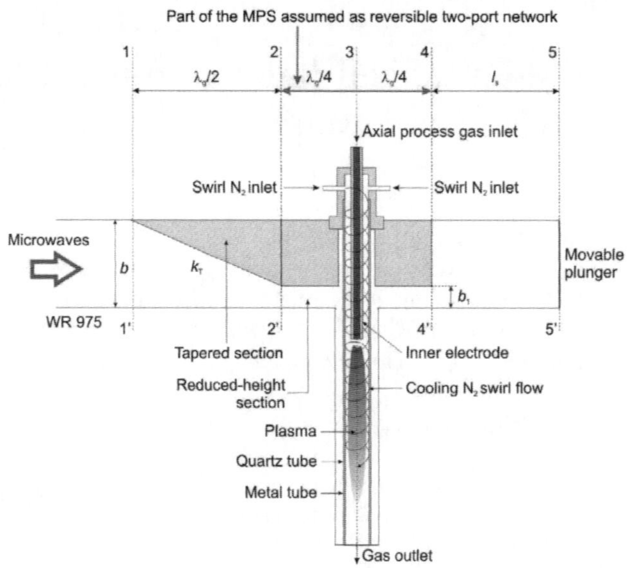

Figure 1. Sketch of the microwave plasma source.

In this MPS, two flows are formed in a quartz tube: central flow (the processed gas) and an additional swirl flow (nitrogen - 50 NL/min). The processed gas flow was introduced to the discharge zone through the inner cylindrical electrode. The additional swirl flow was supplied tangentially through the four inlets, creating a spiral vortex flow in the quartz tube. The swirl flow stabilized the discharge in the center of the quartz tube and protected the quartz wall from overheating [1].

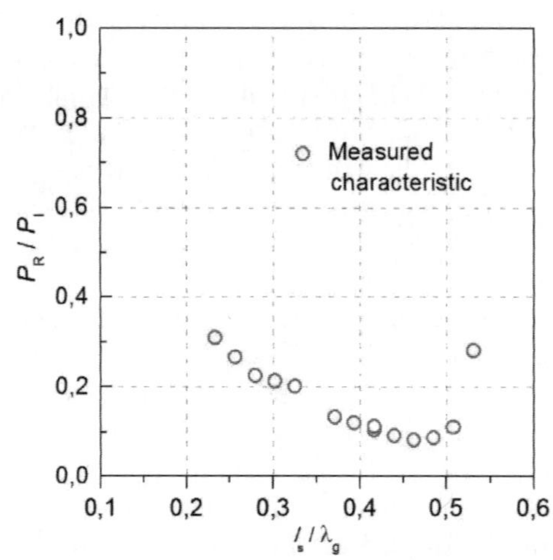

Figure 2. Measured tuning characteristics of the MPS

The experimental setup was described in [4]. The experiment was performed with the nitrogen as process gas with flow rate Q equal to 100 NL/min and incident microwave power P_I = 3 kW. For these working conditions the tuning characteristics of the MPS was measured. The measured characteristics shown in Figure 2 has a minimum at $l_S/\lambda_g \approx 0.46$. This means that at this position of the movable plunger, the efficiency of power transfer from electric field to the plasma in the MPS is highest.

3. Equivalent circuit of the microwave plasma source

Rigorous analysis of behaviour of microwave plasma sources requires solving the Maxwell equations, which is difficult for the microwave structures formed of waveguides and tubes with plasma. However, some properties of the MPSs can be found by using the electric circuit analysis base on the concept of equivalent electric circuit of the MPS [2, 5]. The circuit allows us to determine the normalized admittance y_{in} at the input plane of the MPS (plane 1-1'). The relation between the tuning characteristics P_R/P_I (l_S/λ_g) and the admittance y_{in} is [5]:

$$\frac{P_R}{P_I}\left(\frac{l_s}{\lambda_g}\right) = \left|\frac{y_{in}-1}{y_{in}+1}\right|^2 . \tag{1}$$

Diagram of the proposed equivalent circuit of the MPS is shown in Figure 3. The presented MPS has a region of discontinuity in term of microwave propagation, which is a result of inserting the coaxial-line into the waveguide. Furthermore, the microwave plasma is generated in the discontinuity region. We propose to solve the problem by the use of the Weissfloch circuit [3].

In this paper we assumed that the part of the MPS between planes 2-2' and 4-4' is a two-port network [6]. This section of the MPS is symmetrical to the axis of the inner electrode which means that the assumed two-port network is reversible [6]. For the point of view of the microwave propagation through this region the microwave plasma can be regarded as a loss structure. The idea of the Weissfloch circuit [3] is to represent the reversible lossy two-port network into two sections. The first section represent the losses of the reversible lossy two-port - serial impedance Z_S and parallel resistance R_T. The remaining part, which expresses the area of waveguide discontinuity is represented by the second section in the form of a lossless two-port network. The parameters Z_S and R_T will be called as *plasma impedance* Z_P. The lossless two-port network can be expressed by three impedances Z_a, Z_b and Z_c in T-type circuit. The values of impedances Z_a, Z_b and Z_c can be determined by the simulation of electromagnetic field distribution inside the MPS in the absence of the plasma [6].

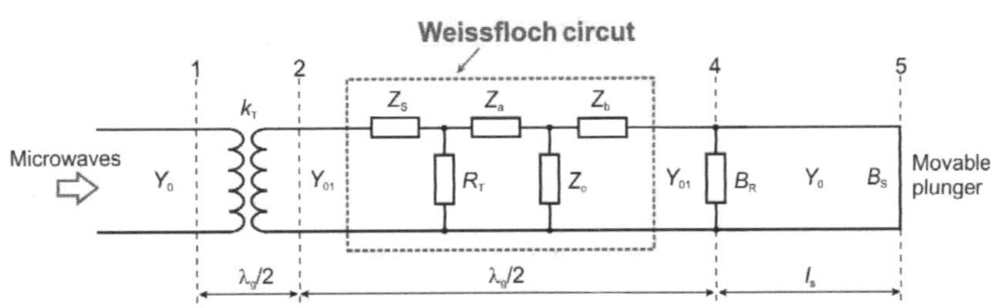

Figure 3. Diagram of the equivalent circuit of the microwave plasma source.

The notation used in figure 3 is as follows:

Y_{01} - characteristic admittance of the waveguide of reduced-height, $Y_{01} = 0.01$ S,
Y_0 - characteristic admittance of the standard WR 975 waveguide, $Y_0 = 0.0025$ S,
$B_S = -\ Y_0\text{ctg}(2\pi\ l_S/\lambda_g)$ - the susceptance introduced by the movable plunger,
B_R - susceptance resulting from abrupt change of height in the MPS waveguide, $B_R = 0.005$ S,
k_{T1} - the transformation factor of the tapered sections, $k_{T1} = Y_0/\ Y_{01}$.

Taking into account the above relation, the normalized input admittance y_{in} takes the form:

$$y_{\text{in}} = \frac{1}{Y_{01}}\left(\frac{1}{\frac{1}{R_T} + \frac{1}{Z_a + \frac{(1 + B_S Z_b + B_R Z_b) Z_c}{1 + B_S (Z_b + Z_c) + B_R (Z_b + Z_c)}}} + Z_S\right)^{-1} \tag{2}$$

As seen from equation (2) the tuning characteristics P_R/P_I (l_S/λ_g) depends on the value of plasma impedance Z_P, which actually is unknown. The compilation of the measured and calculated tuning characteristics may result in finding the unknown electric parameters Z_P of the equivalent circuit.

4. Results and calculations

A three-dimensional simulation of the electromagnetic field distributions inside the part of the MPS were performed using Comsol Multiphysics software [7]. The performed simulation allowed to obtain the impedances Z_a, Z_b and Z_c as follow -j1.6 Ω, -j1.6 Ω and –j793.5 Ω, respectively.

Comparing the calculated tuning characteristics with that measured experimentally a good agreement has been found (Figure 4). The best fitting was obtained for parameters of the plasma impedance Z_P equal to Z_S = (69 - j271) Ω and R_T = 700 Ω.

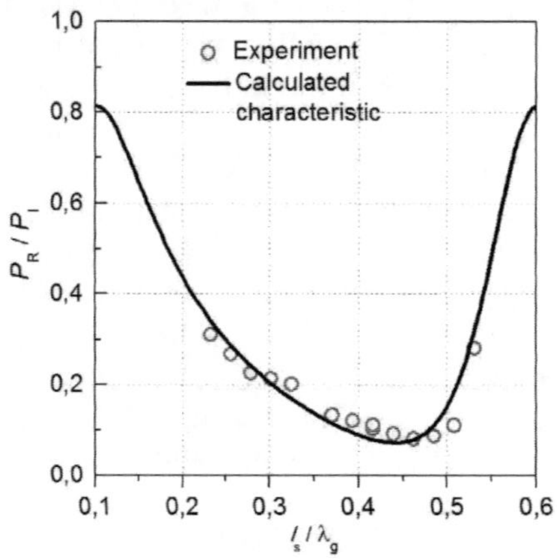

Figure 4. Comparison of measured and calculated tuning characteristics of the MPS.

The equivalent circuit allows to calculate the effect of a metallic diaphragm introduced in the input plane of the MPS which can lead to decrease the ratio of P_R/P_I. Determined reactance of plasma impedance parameter Z_S is capacitive (Im(Z_S) < 0). This indicates that in order to minimize the ratio of P_R/P_I a compensate induction diaphragm should be introduced to the MPS supplied waveguide. In this paper we analyze effect of the symmetrical induction diaphragm [5, 6] on the calculated tuning characteristics of the MPS. Using the presented equivalent circuit the most optimal gap d of the diaphragm and its positions were calculated. The criterion selected in the calculation was to provide a minimum ratio of P_R/P_I to the full extent of the movable plunger position. Sketch and diagram showing location of the diaphragm in the MPS input cross-section are presented in Figure 5, where B_D is susceptance induced in the waveguide by introducing the diaphragm [5, 6].

The calculations showed that the most efficient microwave energy transfer in the MPS was obtained for the gap width d = 150 mm and the distance L = 115 mm (Figure 6). In the calculations the estimated value of plasma impedance Z_P was used. The presented results of calculations allowed to increase efficiency of power transfer from electric field to the plasma, however, experimental confirmation must be performed.

Figure 5. Symmetrical induction diaphragm: a) sketch, b) diagram of location to the MPS input plane.

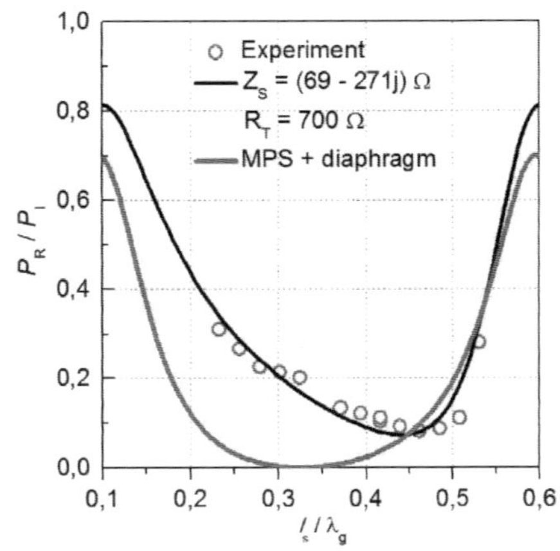

Figure 6. Measured and calculated tuning characteristics of the MPS including the diaphragm.

5. Conclusions

This work presents a version of the equivalent circuit correctly describing the properties of the MPS. A lossy two-port network, as countertype of the area of waveguide discontinuity with the discharge, was used. The proposed concept of the equivalent circuit of the MPS has a versatile properties and can be applied to other MPS structures.

The MPS presented in this paper showed a stable operation, although, optimization of this device is required. The presented investigations are the first step in optimization of the MPS. The aim of the next step will be finding, by numerical modeling, an optimal dimensions of the structural elements of this device which will ensure the minimum ratio P_R/P_I in the wide range of the movable plunger position.

Acknowledgments

We are grateful to The National Science Centre (programme no. 2013/11/N/ST8/00802) for the financial support of this work.

References

[1] Mizeraczyk J, Jasiński M, Nowakowska H and Dors M 2012 *Nukleonika* **57** (2) 241
[2] 1992 *Plasma technology, 4 Microwave Excited Plasmas* ed M Moisan and J Pelletier (Amsterdam: Elsevier) chapter 2 pp 11–52
[3] Weissfloch A 1942 *Hochfrequenz Tech.* **60** 67
[4] Hrycak B, Czylkowski D, Jasiński M and Mizeraczyk J 2012 *Electrical Review* **11b** (88) 310
[5] Sobański M, Lubański M, Jasiński M and Mizeraczyk J 2013 *Electrical Review* **5** (89) 254
[6] Panecki M, Litwin R and Drozdowicz L 1961 *Teoria i Technika Mikrofalowa* ed B Paszkowski and T Gawron (Warszawa: Wydawnictwo Naukowo Techniczne) chapter 2 pp 34–56
[7] http://www.comsol.com

IOP Conf. Series: Materials Science and Engineering **113** (2016) 012009 doi:10.1088/1757-899X/113/1/012009

Magnetic thin film deposition with pulsed magnetron sputtering: deposition rate and film thickness distribution

M Ozimek[1, 2], W Wilczyński[3], B Szubzda[1]
[1]Electrotechnical Institute Division of Electrotechnology and Materials Science, Wroclaw, Poland
[2]International Laboratory of High Magnetic Fields and Low Temperatures, Wroclaw, Poland
[3]Electrotechnical Institute, Warsaw, Poland

E-mail: mariusz.ozimek@ml.pan.wroc.pl

Abstract. The goal of conducted study was an experimental determining the relations between technological parameters of magnetron sputtering process on deposition rate (*R*) and thickness uniformity of magnetic thin films. Planar $Ni_{79}Fe_{16}Mo_5$ target with a diameter of 100 mm was sputtered in argon (Ar) atmosphere. Deposition rate was measured in a function of gas pressure, target power and target-substrate distance. The highest value of $R \approx 280$ nm·min^{-1}. The obtained results in deposition rate of magnetic film were compared to deposition rate of cooper (Cu), aluminum (Al), titanium (Ti) and titanium oxide (TiO_x) and the deposition rate of Ni-Fe alloy were higher that Al and Ti. The film thickness distribution was measured for radial distance from the target centre ranging up to 60 mm and target-substrate distance ranging form 70 to 115 mm. Among others it was stated that for the larger value of target-substrate distance the larger uniform of film thickness are obtained.

1. Introduction

Magnetron sputtering is one of well-developed method for thin film fabrication but its extensive use in industrial application depends on the ability of obtaining high quality films with high value of deposition rate on a large deposition area. In case of large deposition are, the film uniformity play a major role and generally for different sputtering systems and different sputtering conditions, the characteristics of the sputtered atoms and the thickness profile are different. Therefore, the manifestation of the relationship between the film uniformity and the deposition parameters is required.

In magnetron technique there is a number of ways to apply the voltage to the cathode depending on the nature of the discharge: AC, DC, Pulse DC, radio frequency etc. Pulsed power operating in mid-frequency range (20-300 kHz) is a relatively new method [1-3]. The goal of this type has been mainly improving the stability of reactive processes, however, currently begins to be increasingly used in high-performance non-reactive processes. Pulsed powering almost completely suppresses the arcing and makes the process more stable, even while maintaining high power densities and the values of obtained deposition rates are even approx. 500 nm·min^{-1} for reactive processes [4, 5] and approx. 5000 nm·min^{-1} for neutral processes [6, 7].

In this article the dependence between the process parameters and the value of deposition rate and thickness profile of magnetic thin film is described. One limitation of magnetron sputtering techniques is the limited ability for using ferromagnetic targets. The main problem is that the magnetic nature of the target material interferes with the magnetic field generated by the magnetron resulting in reducing the efficiency of the process and bad film homogeneity. The iron-nickel alloy (called Superpermalloy was used as a target. The choice of this material was dictated by its soft magnetic properties ($\mu >> 1$) and wide applications, especially in modern technology. The obtained results in deposition rate were compared to deposition rate of cooper (Cu), aluminum (Al), titanium (Ti) and titanium oxide (TiO_x) thin films.

2. Experimental

The investigation was performed in a vacuum system (0,14 m^3) equipped with rotary and turbomolecular pumps. The final pressure of the deposition chamber was about $6 \cdot 10^{-4}$ Pa.
The deposition rate (R) was measured as a function of technological parameters of magnetron sputtering process, by using the SQM-160 quartz crystal monitor.
Thickness profile was examined by means of atomic force microscope (AFM), by measuring the high difference between the surface of the substrate and thin film surfaces (Figure 1). In this case magnetic thin films were deposited on the Corning glass (20×20 mm) and the shutter was used to shield the glass plate from deposition, while the sputtering parameters were set and the target was conditioned.
In all cases Ni-Fe alloy thin films were obtained by using $Ni_{79}Fe_{16}Mo_5$ target with 100 mm in diameter and 1 mm thick. The cathode was sputtered in pure argon (99,999 %Ar) atmosphere at pressures (p_{Ar}) ranged from 0,4 to 5,0 Pa. Magnetron target was powered by DPS Power Supply (medium frequency power supply) unit [4] with target power (P) up to 2000 W. The target-substrate distance ($d_{T\text{-}S}$) was changed from 70 to 115 cm. The film thickness distribution was measured for radial distance (a) from the target centre ranging from 0 (the magnetron centre) to 60 mm.
To compare the deposition rate of NiFe alloy film with others deposited materials the Cu (99,999 %Cu), Al (99,999 %Al) and Ti (99,999 %Ti) targets with 100 mm in diameter and 3 mm thick were used. All targets were sputtered in the same condition as $Ni_{79}Fe_{16}Mo_5$. The Ti target was in addition sputtered in pure oxygen (99,999 %O_2) atmosphere.

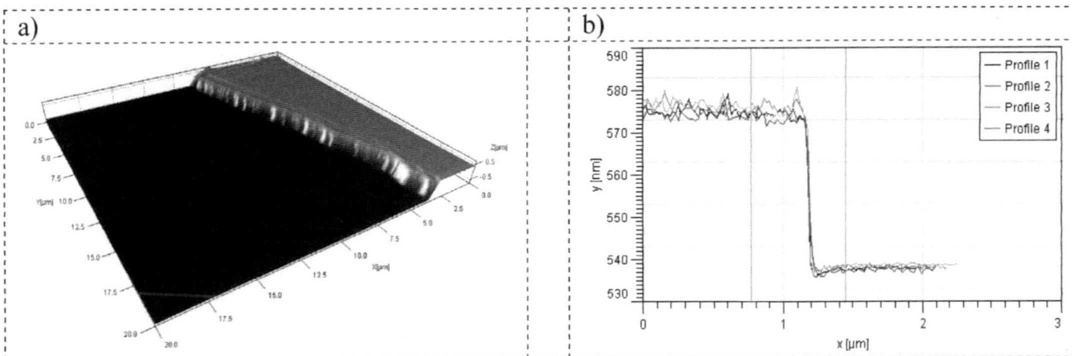

Figure 1. The AFM image of magnetic thin film cross-section (a) and result of measuring its high (b).

3. Results and discussion

3.1. Deposition rate

3.1.1. Gas pressure effect

Deposition rate of thin magnetic film in dependence on gas pressure is shown in figure 2. The results are referred to the experiment carried out in the pressure ranged from 0.4 Pa to 5.0 Pa and target-substrate distance $d_{T\text{-}S}$=70 and 115 mm. The applied target power P were changed in the range from 400 to 2000 W. Based on the characteristics it can be seen that for pressures between 0.4 and 1.0 Pa, increase in pressure only cause a slight change in the deposition rate. At pressures above 1.0 Pa, increase in pressure results in decrease in the deposition rate. This, of course, is as would be expected. As the sputtered particles travel between the target and the substrate, they collide with the gas atoms and are scattered. This process can be characterized by the mean free path λ that is related among others to the gas pressure. As mentioned in many studies, increasing pressure increases gas scattering, and hence, decrease the probability of a sputtered atom depositing on the substrate [8-11]. This effect is particularly significant at pressures higher than 1.0 Pa [12, 13].
The obtained results showed that the values of R at target-substrate distance $d_{T\text{-}S}$=115 mm are similar but almost twice as small as at $d_{T\text{-}S}$=70 mm. It is reasonable to assume that the bigger the target-substrate distance, the thinner thin films are obtained.

a) b)

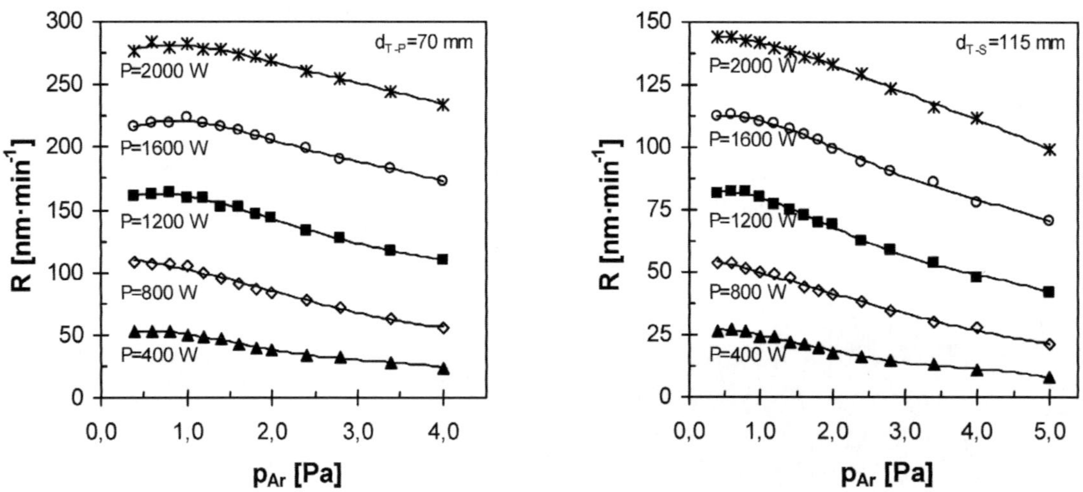

Figure 2. Deposition rate R of Ni-Fe alloy thin film in dependence on the argon pressure p_{Ar} in target-substrate distances $d_{T\text{-}S}$=70 mm (a) and $d_{T\text{-}S}$=115 mm (b).

3.1.2. Target power effect

Figure 3 shows the Ni-Fe alloy thin film deposition rate at several sputtering process parameters. As expected, an increase in target power results in an increase in the deposition rate.
In our study the target power increase was generated by the increase of filling of the sputtering pulse. It causes increase in an effective sputtering time. Finally increase in sputtering time causes: i) the increase in the sputtered particle flux density and ii) pressure reduction in the region near the magnetron cathode, due to energy dissipation from sputter-ejected atoms within the gas. In the second case it causes a certain rarefaction of discharge gas near the target and consequent increase of the

mean free path of sputtered atoms. In the examined range of P the R changed linearly with the applied power.

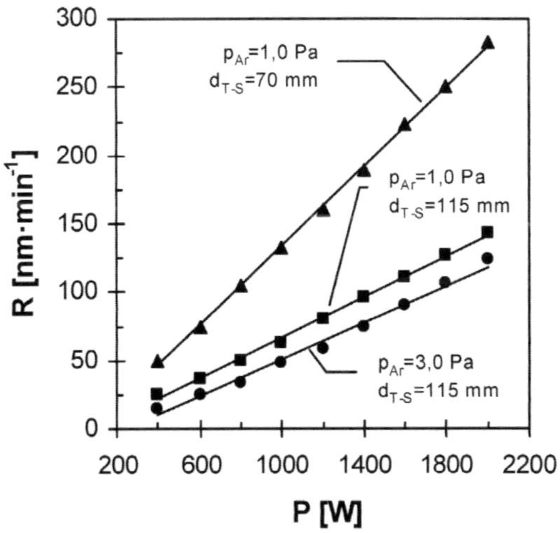

Figure 3. Deposition rate R of Ni-Fe alloy thin films in dependence on the target power P.

In comparison to other thin film materials, the obtained results in deposition rate for magnetic film are even higher than for aluminum and titanium (Figure 4). As can be seen the highest values of the deposition rates obtained for copper (Cu). The lowest deposition rate was measured for titanium oxide, but it is understandable that effectiveness of reactive processes is lower than nonreactive, due to the target surface oxidation. For P=2000 W, the R values are as follows: $R_{Cu}\approx 290$ nm·min^{-1}, $R_{NI\text{-}Fe}\approx 140$ nm·min^{-1}, $R_{Al}\approx 110$ nm·min^{-1}, $R_{Al}\approx 45$ nm·min^{-1} and $R_{TiOx}\approx 5$ nm·min^{-1} The resulting dependence R=f(P), in the measured target power range are linear.

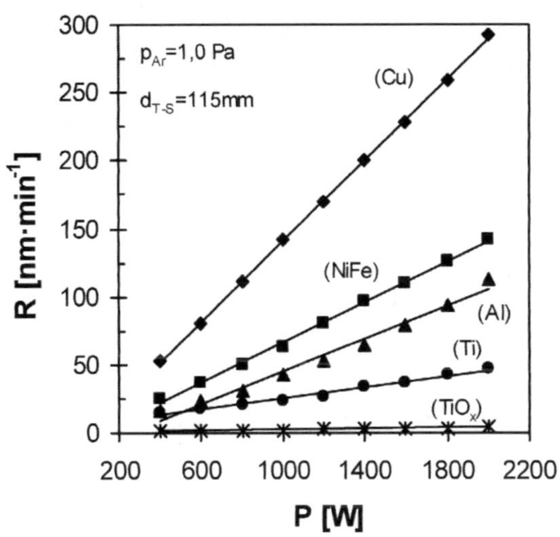

Figure 4. Deposition rate R of various thin films materials in dependence on the target power P.

3.2. Film thickness distribution

In figure 5, the Ni-Fe alloy film thickness distribution, plotted against radial distance a from the magnetron center for several values of the target-substrate distance and target power. The obtained results showed that in the studied range of $d_{T\text{-}S}$ and a, the thicknesses of prepared film decrease with its

IOP Conf. Series: Materials Science and Engineering **113** (2016) 012009 doi:10.1088/1757-899X/113/1/012009

increasing, but for the larger target-substrate distance the larger uniform of film thickness were obtained. Generally, the sputtering gas pressure also affects the film thickness distribution although, in our case, the effect is not very obvious. Further study is still required to clarify these effects.

The thickness decreasing in a function of *a* was confirm by other studies. As can be read, the thickness distribution of thin film depends not only on the process parameters but is also conditioned by a magnetron construction (magnetron geometry and its magnetic field) [14-17].

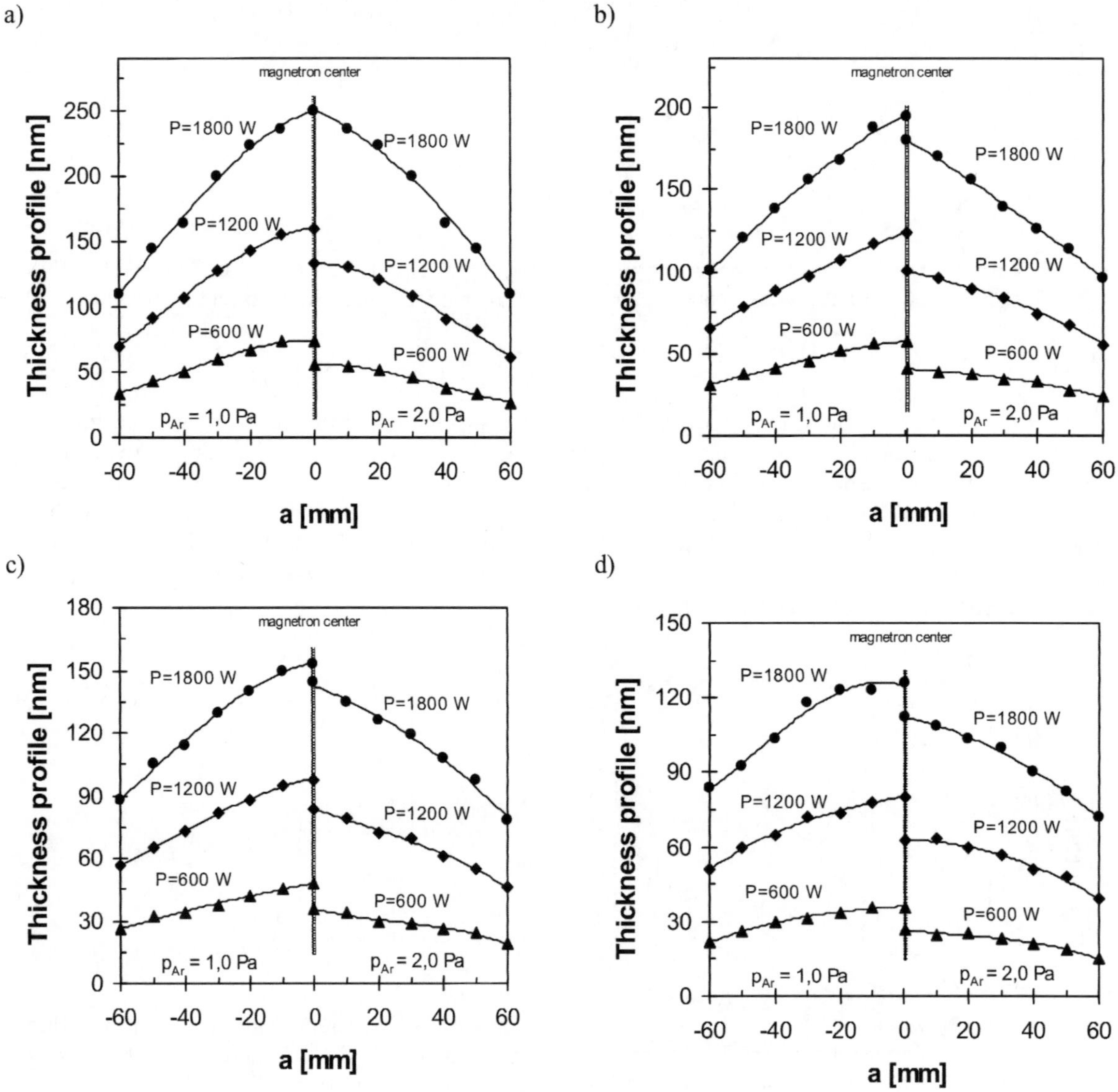

Figure 5. Thickness profile of Ni-Fe alloy thin films measured for different target power *P* and substrate-distance of 70 mm (a), 85 mm (b), 100 mm (c) and 115 mm (d). Argon pressure p_{Ar}=1.0 Pa (left side) and 2.0 Pa (right side).

IOP Conf. Series: Materials Science and Engineering **113** (2016) 012009 doi:10.1088/1757-899X/113/1/012009

4. Conclusions

The influence of magnetron process parameter on deposition rate *R* and thickness profile of magnetic thin film, by sputtering $Ni_{79}Fe_{16}Mo_5$ target, has been presented. The obtained results can be summarized as follows:

(i) In case of gas pressure effect: above 1.0 Pa, deposition rate are inversely proportional to pressure. Up to the 1.0 Pa the values of deposition rate are almost on the same level.

(ii) The variation of Ni-Fe ally deposition rate with target power, at the range 400-2000 W gives evidence of a linear dependence of *R* with *P*.

(iii) The highest value of $R \approx 280$ nm·min^{-1} obtained for P=2000 W and $d_{T\text{-}S}$=70 mm. Our further work revealed that the values of deposition rate of Ni-Fe alloy thin film is higher than Al and Ti.

(iv) One of the factor which influence the film thickness uniformity is the variation of the target-substrate distance. Another is the gas pressure. In our investigation for larger $d_{T\text{-}S}$ the more uniform films were obtained. In case of p_{Ar} on thickness profile the further study are required.

5. References

[1] Kelly P J and Bradley J W 2009 *J. Optoelectronics and Adv. Mat.* **11(9)** 1101.
[2] Ziaja J 2007 *Przegląd Elektrotechniczny* **11** 235.
[3] Arnell R D, Kelly P J and Bradley J W 2004 *Surface & Coatings Technology* **188–189** 158.
[4] Posadwoski W M, Wiatrowski A, Dora J and Radzimski Z J 2008 *Thin Solid Films* **516** 4478.
[5] Posadowski W M 1999 *Thin Solid Films* **343-344** 85.
[6] Brudnik A, Czapla A and Posadowski W M 2008 *Vacuum* **82** 1124.
[7] Posadowski W M 2001 *Thin Solid Films* **392** 201.
[8] Ekpe S D, Bezuidenhout L W and Dew S K 2005 *Thin Solid Films* **474** 330.
[9] Helmer J C and Wickersham C E 1986 *J. Vac. Sci. Technol. A* **4(3)** 408.
[10] Ekpe A D and Dew S K 2002 *J. Vac. Sci. Technol. A* **20(6)** 1877.
[11] Valles-Abarca J A and Gras-Marti A 1984 *J. Appl. Phys.* **55(5)** 1370.
[12] Drusedau T P., Lohmann M. and Gerke B. 1988 *J. Vac. Sci. Techol. A* **16(4)** 2728.
[13] Escrivao M L, Pereira P J S, Ferreira J L, Teixeira M R and Maneira M J P 2002 *Vacuum* **64** 367.
[14] Hong S, Kim E, Bae B, No K, Lim S, Woo S and Koh Y 1996 *J. Vac. Sci. Technol. A* **14(5)** 2721.
[15] Fu C, Yang C, Han L and Chen 2006 *Surface & Coatings Technology* **200** 3687.
[16] Blech I A and Vander Plas H A 1983 *J. Appl. Phys.* **54(6)** 3489.
[17] Fan Q, Chen X and Zhang Y 1995 *Vacuum* **46(3)** 229.

Acknowledgments

The research was co-supported by Electrotechnical Institute Division of Electrotechnology and Materials Science and Wroclaw Research Center EIT+ under the project "The Application of Nanotechnology in Advanced Materials" – NanoMat (POIG.01.01.02-02-002/08) financed from the European Regional Development Fund (Operational Programme Innovative Economy, 1.1.2).

EMI shielding using composite materials with two sources magnetron sputtering

J Ziaja, M Jaroszewski, M Lewandowski
Department of Electrical Engineering Fundamentals,
Wroclaw University of Technology, Poland

E-mail: marcin.p.lewandowski@pwr.edu.pl

Abstract. In this study, the preparation composite materials for electromagnetic shields using two sources magnetron sputtering DC-M is presented. A composite material was prepared by coating a nonwoven polypropylene metallic layer in sputtering process of targets Ti (purity 99%) and brass alloy MO_{58} (58%Cu, 40%Zn, 2%Pb) and ϕ diameter targets = 50 mm, under argon atmosphere. The system with magnetron sputtering sources was powered using switch-mode power supply DPS (Dora Power System) with a maximum power of 16 kW and a maximum voltage of 1.2 kV with group frequency from 50 Hz to 5 kHz. The influence of sputtering time of individual targets on the value of the EM field attenuation *SE* [dB] was investigated for the following supply conditions: pressure $p_p = 2\text{x}10^{-3}$ Torr, sputtering power P = 750 W, the time of applying a layer t = 5 min, group frequency f_g = 2 kHz, the frequency of switching between targets f_p = 1 Hz.

1. Introduction

Due to the rapid increase of sources generating the electromagnetic fields (radio broadcasting, television, radio communication, mobile networks) and power generated by PEM sources, the shielding design is getting more and more challenging. Materials used in the technique of EMI shielding must have a suitably high coefficient of the shielding effectiveness *SE*, be resistant to influence of external environment (corrosion, oxidation) and mechanical impact. Materials must be also easy to form the shield and must have low production costs. Shields made as metal sheets, foil and mesh are characterised by good EM filed shielding effectiveness coefficient. However they are characterised by low resistance to influence of external environment (especially corrosion) and relatively high weight. The alternative to metal shields is the use of composite shield. The engineering of materials used for electromagnetic shielding is one of the most widely developing field of applications of composite materials [1-3]. One of the group composite materials based on laminar systems. Typical laminar system consists of a nonwoven fabrics with conductive layers deposited or laminates with pressed in conductive layers. Conductive layers may be put on by using the following methods: silk-screen printing [4], vacuum evaporation, or magnetron sputtering [5-7]. They are characterised by high shielding efficiency coefficient. However not every materials is suitable for conductive coating. The most often used method of the obtaining conductive layers is magnetron sputtering. Additionally during a process of magnetron sputtering, plasma cleans and activates surface of substrates [8,9]. Therefore magnetron methods allow to do metallization of materials such as PTFE or PP, which can not by transformed by metallization with other methods, because of their surface properties. Previous studies have shown that metallic layers (Zn) deposited on polypropylene nonwoven fabric allow effective electric field shielding (above 50 dB) [10]. To increase the resistant to influence of external environment and mechanical impact the two sources magnetron sputtering was proposed.

2. The two sources magnetron sputtering system

The processes of sputtering of two sources (targets): titanium (purity 99%) and brass alloy (58%Cu, 40%Zn, 2%Pb) were realised by magnetron launchers WMK-50 with the magnetic system of diameter ϕ=50 mm. The argon was used as a working gas. The pressure in the vacuum chamber was $p_p = 2x10^{-3}$ Torr. The distance between the target and nonwoven fabrics has been set to 100 mm. The magnetron launcher was powered by impulse current source type DPS (Dora Power System) witch works in DC-M mode controlling power up to 16 kW with a maximum voltage up to 1.2 kV. This power supply can control group frequency f_g in the range of 50Hz to 5kHz and frequency of switching between targets f_p in the range of 0,1 Hz to 50 Hz.

3. Composites materials PP/brass alloy/Ti

Titanium and brass alloy layers were deposited on polypropylene nonwoven fabric of the base weight of 180g/m^2. The power P emitted by the gun was 750 W and the time of depositing t was 5 minutes. The frequency of switching between targets f_p was 1Hz. The PP nonwoven fabric were deposited only on A side or B side (5 samples – A side, 5 samples B side) for the different ratios of the time sputtering targets of titanium and alloy brass. A side was significantly less expanded surface area relative to the B side. The values of the surface resistivity of the coated samples are shown in Table 1. In figure 1 and figure 2 nonwoven fabric surface (sample A4 and B4) after the magnetron sputtering process were shown.

Table 1. Composites materials PP/brass alloy/Ti.

	Titanium (%)	Brass alloy (%)	Surface resistivity ρ_s (Ω)
A1	100	0	160k
A2	75	25	3,2k
A3	50	50	700
A4	25	75	270
A5	0	100	250
B1	100	0	4,9k
B2	75	25	1k
B3	50	50	98
B4	25	75	33
B5	0	100	24

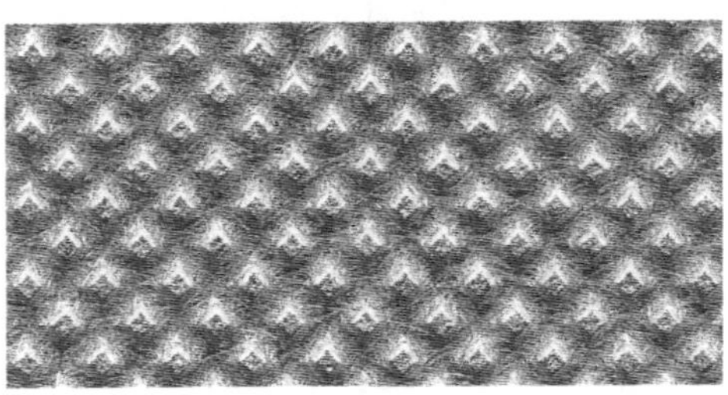

Figure 1. Textile surface after deposition of the brass alloy (75%) and titanium (25%) layer. Coated A site – A4 sample.

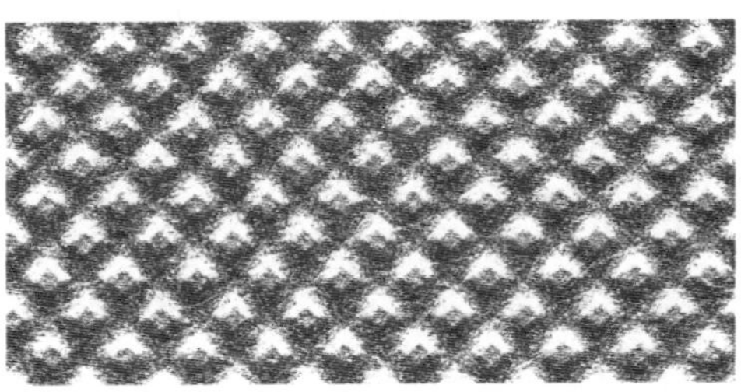

Figure 2. Textile surface after deposition of the brass alloy (75%) and titanium (25%) layer. Coated B site – B4 sample.

4. Shielding properties of PP/brass alloy/Ti composites.

Measurements of shielding attenuation were realised on the test setup prepared in the Institute of Telecommunications, Teleinformatics and Acoustics of Wrocław University of Technology in accordance with the method of ASTM D4935-99. Measuring uncertainty was approximately 2 dB.
The study have shown an influence of different ratios of the time sputtering targets of titanium and alloy brass on a value of shielding attenuation. The highest value of SE was obtained for nonwoven fabric coated only by brass alloy (sample A5 and B5). It should be noted that the addition of titanium below 25% did not decrease value of *SE* (sample A4 and B4). A significant impact on the value of *SE* had surface structure. B side was more expanded surface, therefore the value of *SE* was almost two times higher. Dependence of the *SE* on frequency electromagnetic field for A side and B side was shown in figure 3 and figure 4. The frequency of electromagnetic field was in the range of 10 MHz to 2 GHz.

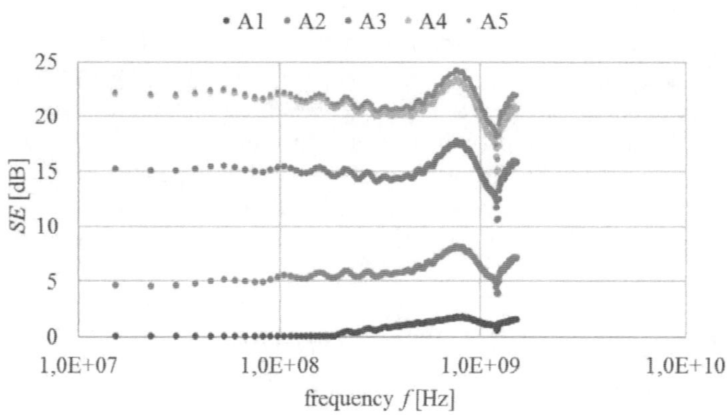

Figure 3. Dependence of the coefficient of shielding efficiency *SE* on frequency *f*. Coated A side.

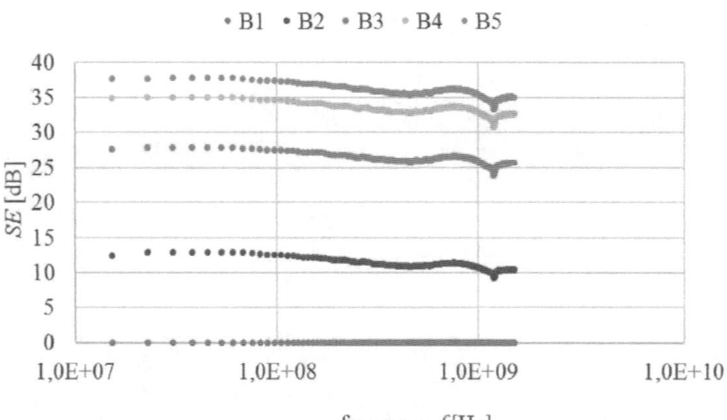

Figure 4. Dependence of the coefficient of shielding efficiency *SE* on frequency *f*. Coated B side.

In figure 5 dependence of the coefficient of shielding efficiency *SE* on surface resistivity ρ_s. was shown. The observed dependence in figure 5 was a linear in a log scale.

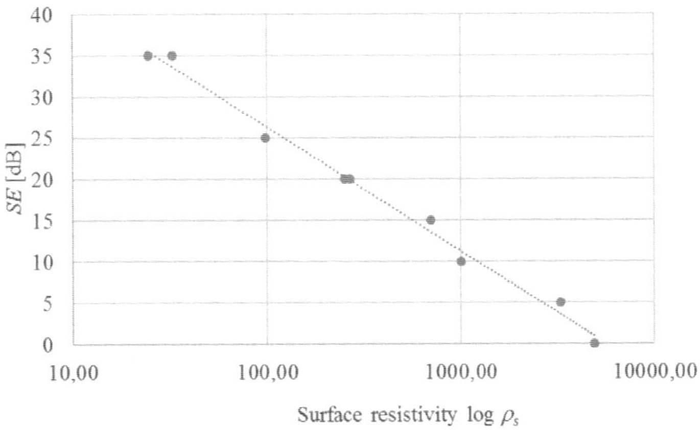

Figure 5. Dependence of the coefficient of shielding efficiency *SE* on surface resistivity ρ_s. Coated A or B side.

5. Conclusion

The possibility of industrial fabrication of the composite shielding materials with the coefficient of shielding efficiency exceeding 40 dB was confirmed by obtained results. The addition of titanium below 25% did not decrease value of *SE*. The addition of Ti could increase the resistant to influence of external environment and mechanical impact. The metallization of polypropylene nonwoven fabric with very good adhesion was realized using magnetron sputtering methods. Using the method of two sources magnetron sputtering allowed to speed up of technological process in the preparation of multilayer structures.

Acknowledgements

The investigations were carried out within the statutory research financed by Ministry of Science and Higher Education.

References

[1] Jaroszewski M, Pospieszna J and Ziaja J **2010** *J. Non-Cryst. Solids* vol. 365(11-17) 625-628
[2] Bula K, Koprowska J and Janukiewicz J **2006** *Fibres & Textiles in EE* vol. 14(5) 75-79
[3] Koprowska J, Pietraniak M and Stawski W **2004** *Fibres & Textiles in EE* vol. 12(3) 39-42
[4] Wang Li-Li, et al. **2009** *Carbon* vol. 47 1905-1910
[5] Ziaja J, **2007** *Electrical Review* vol. 83(11) 235-239
[6] Ziaja J, Ozimek M and Janukiewicz J **2010** *Electrical Review* vol. 86(5) 222-224
[7] Ozimek M, Sikora A, Wilczyński W The study of magnetic domain development in shielding films using MFM technique, *Przegląd Elektrotechniczny* 89(8) **2013** 208-210
[8] Szubzda B, Szmaja A, Ozimek M and Mazurkiewicz S **2014** *Applied Physics A* vol. 117(4) 1801-1809
[9] Żyłka P **2012** *Electrical Review* vol. 88(5) 128-131
[10] Ziaja J, Koprowska J and Janukiewicz J **2008** *Fibres & Textiles in EE* vol. 16(5) 64-66

IOP Conf. Series: Materials Science and Engineering **113** (2016) 012011 doi:10.1088/1757-899X/113/1/012011

Design, fabrication and characterization of LTCC-based electromagnetic microgenerators

M Gierczak[1], P Markowski and A Dziedzic
Wrocław University of Technology, Faculty of Microsystem Electronics and Photonics, Wybrzeże Wyspiańskiego 27, 50-370 Wrocław, Poland

E-mail: miroslaw.gierczak@pwr.edu.pl

Abstract. Design, manufacturing process and properties of electromagnetic microgenerators fabricated in LTCC (Low Temperature Co-fired Ceramics) technology are presented in this paper. Electromagnetic microgenerators consist of planar coils spatially arranged on several layers of LTCC and of a multipole permanent magnet. Two different patterns of coils with 2-, 8-,10- and 12-layers and outer diameter of 50 mm were designed and fabricated. Silver-based pastes ESL 903-A or DuPont 6145 were used. In order to estimate the inductance of a single spatial coil the Greenhouse (self-inductance) and Hoer (mutual inductance) calculation methods were used. To verify the calculation results a single-layer coil was fabricated for each pattern and its inductance was measured using the precision RLC Meter. Fabricated LTCC microgenerators with embedded coils allow to generate voltage higher than ten volts and the electrical output power of approximately 600 mW at the rotor rotation speed of 12 thousands rpm. The self-made system was used for characterization of LTCC-based electromagnetic microgenerators.

1. Introduction

The paper focuses on design, fabrication and characterization of the electromagnetic microgenerators consisted of planar coils spatially arranged on several LTCC layers and a multipole permanent magnet. Twelve different electromagnetic generators consisted of 2-, 8-, 10- and 12-layers were fabricated using LTCC technology. 28 coils were screen-printed on each layer. Two different layers and silver-based pastes ESL 903-A or DuPont 6145 were used. Diameter of each generator was 50 mm.

Greenhouse [1] (self-inductance) and Hoer [2] (mutual inductance) calculation methods were used to estimate the inductance of a single spatial coil.

The measurement system for characterization of fabricated electromagnetic microgenerators was designed and created. To generate alternating magnetic field a 28-pole permanent magnet was mounted on rotor of three-phase Brushless Direct Current (BLDC) motor. Measurement system included also driver for that type of motor and tachometer [3].

[1] To whom any correspondence should be addressed.

IOP Conf. Series: Materials Science and Engineering **113** (2016) 012011 doi:10.1088/1757-899X/113/1/012011

2. Design and fabrication of electromagnetic microgenerators

LTCC technology allows the fabrication of quite small generators. There were fabricated twelve different types of structures. Each was consisted of several layers of coils electrically connected (trough via holes) in one circuit.

The masks for two different patterns of coils are shown in Figs. 1 and 2. There are 28 unconnected coils with connection pads. The single coils are shown in Fig. 3. Width of screen-printed paths was equal to 200 μm, distance between each twists - about 150 μm. To connect the coils two additional layers - top and bottom – were added to connect the coils in series (Fig. 4). The way of connecting the individual layers in 2-, 8-, 10- and 12-layer generators is shown in Fig. 5.

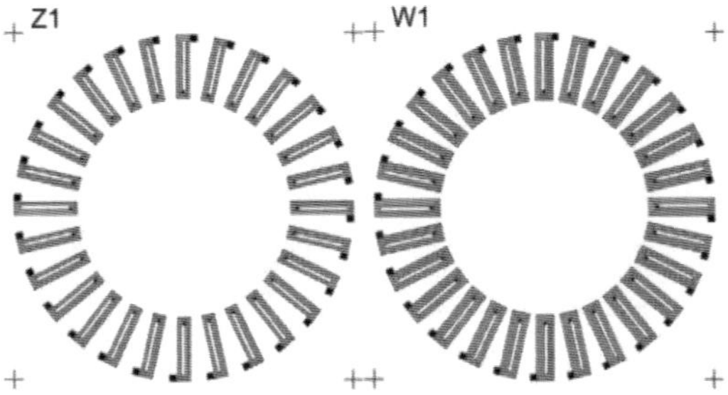

Figure 1. Mask's design of first pattern – 28 coils on the single layer.

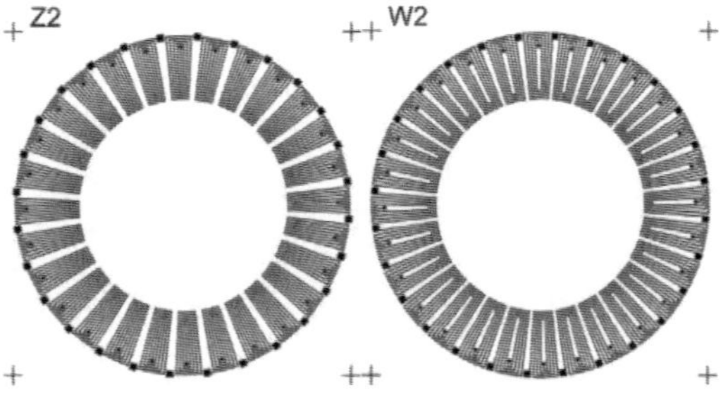

Figure 2. Mask's design of second pattern – 28 coils on the single layer.

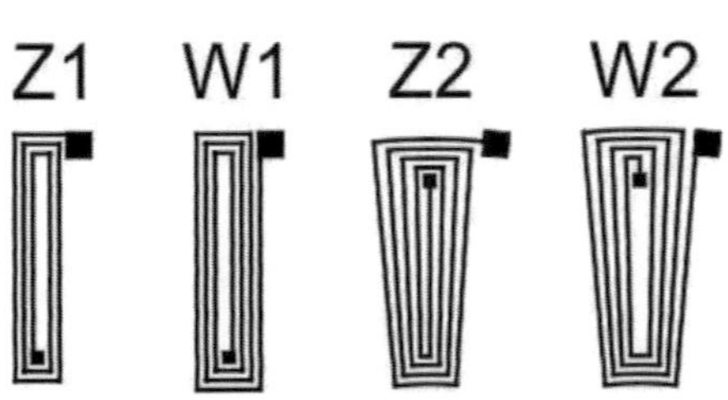

Figure 3. Mask's design – one coil of every pattern.

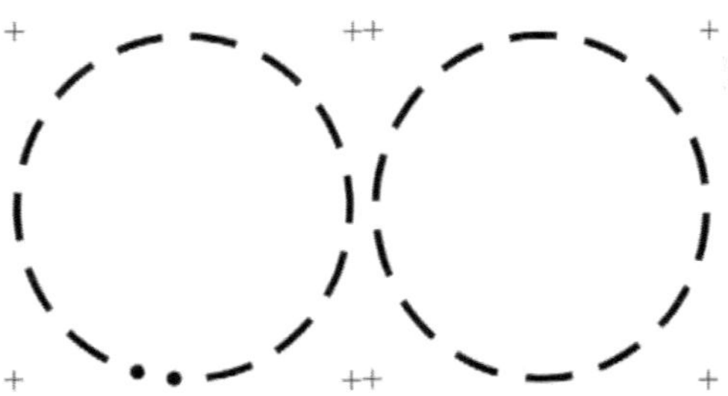

Figure 4. Mask's design – the top and bottom layer.

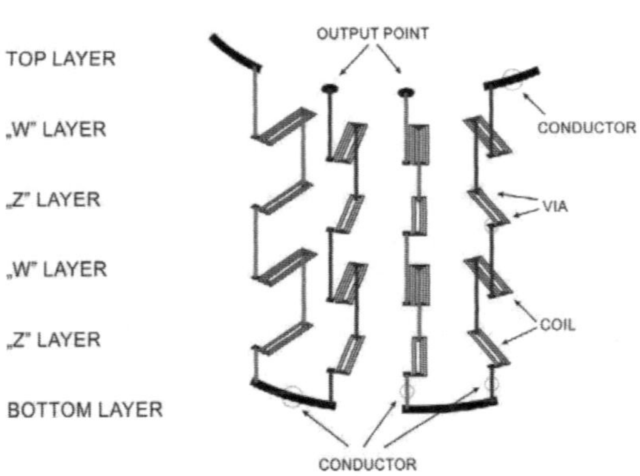

Figure 5. The way of connecting the individual layers.

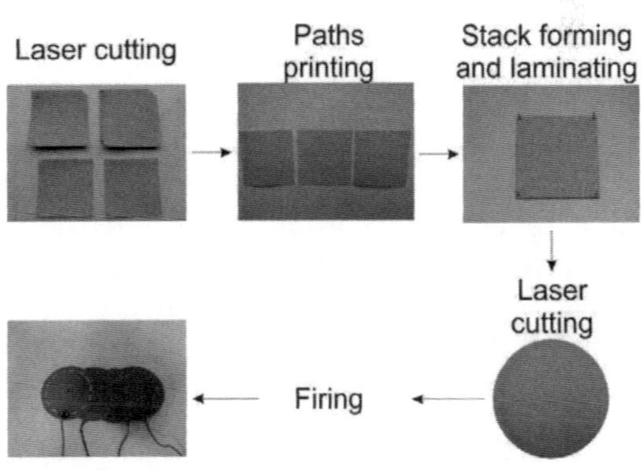

Figure 6. The fabrication steps of the electromagnetic LTCC generators.

The generators' fabrication steps are shown in Fig. 6. Firstly, unfired LTCC sheets (green tape) were cut to the specific size. After screen printing of coils and connection paths (DuPont or ESL, silver-

IOP Conf. Series: Materials Science and Engineering **113** (2016) 012011 doi:10.1088/1757-899X/113/1/012011

based ink) proper number of layers were put together and laminated at 200 atm. In the next step the structures were trimmed to a circular shape and fired in the time/temperature profile typical for the LTCC technique [4].

3. The electromagnetic microgenerators' characterization system

The measurement system for testing the electromagnetic generators, presented previously in [3] was modified. New rotating disc with 28 magnets was designed [5]. The disk was fabricated using 3D-printer. Magnets were pressed into holes, than the disc was mounted on the motor's rotor and on the measuring table (Fig. 7). Fabricated coils were screwed to the thread located centrally on the motor's stator. The distance between the coils and magnets was 3.05 mm and it was provided by a spacer made in LTCC technology (Fig. 8).

Figure 7. The rotating disc with 28 magnets and the spacer fabricated in LTCC technology.

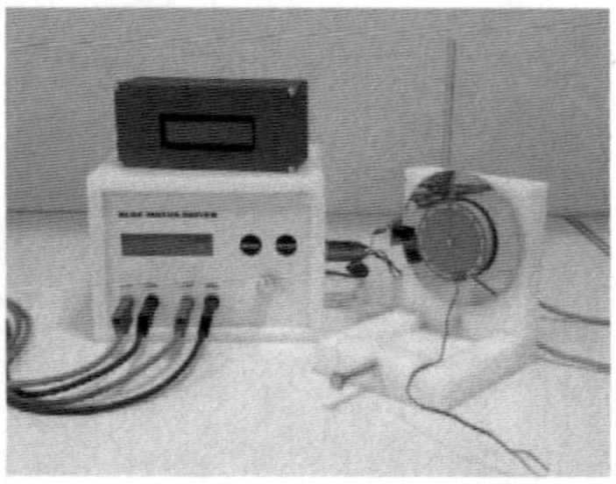

Figure 8. The electromagnetic microgenerators' characterization system.

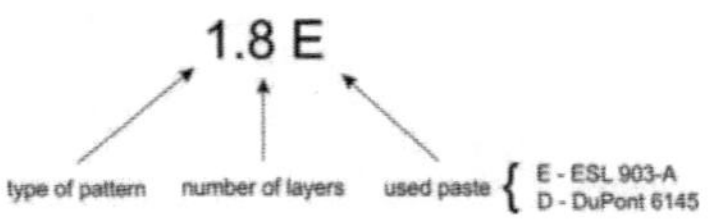

Figure 9. The code scheme used to distinguish the structures.

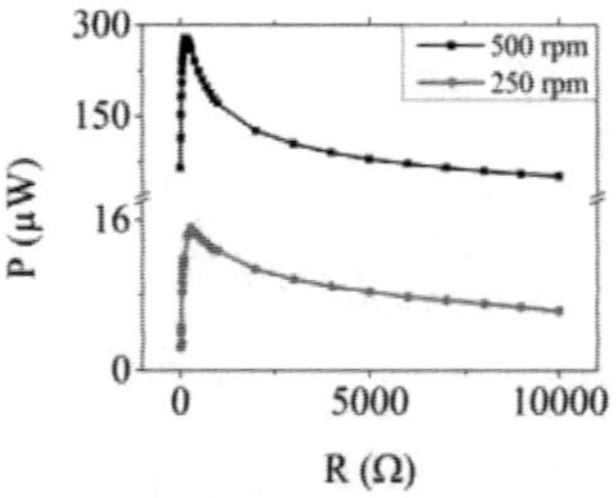

Figure 10. The output power *vs* load resistance (2.8 D).

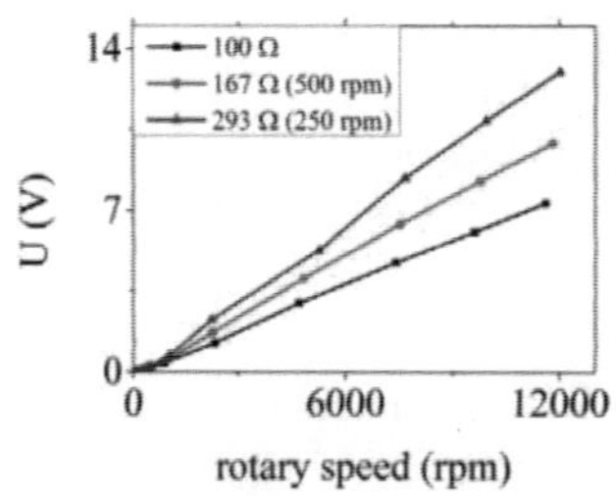

Figure 11. The voltage *vs* rotary speed (2.8 D).

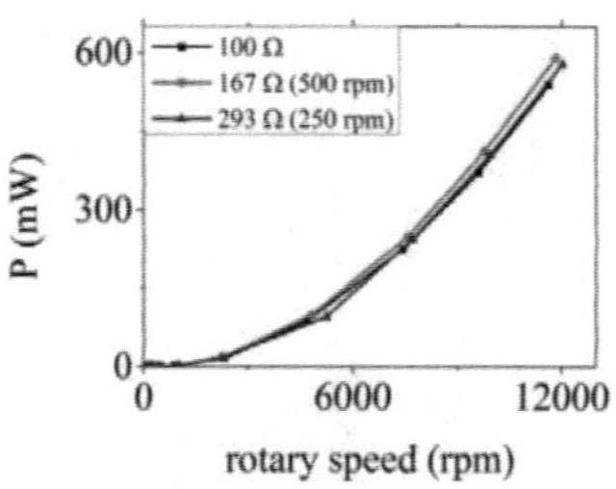

Figure 12. The output power *vs* rotary speed (2.8 D).

The code scheme used to distinguish different structures was explained in Fig. 9.

For each generator the load resistance for two speeds: 250 rpm and 500 rpm was determined (eg. Fig. 10 – load resistance 293 Ω and 167 Ω, respectively). For determined operating points the characteristics of voltage *vs* rotary speed (Fig. 11) and output power *vs* rotary speed (Fig. 12) were measured. Moreover, in order to compare all generators, such characteristics were also measured for the load 100 Ω. During the measurements the rheostat decade was used as the load whereas the voltage was measured by Fluke 8842 multimeter. The voltage delivered by the generator was rectified using a simple circuit with four Schottky diodes (1N5819) and a 1 μF smoothing capacitor.

4. Results

To analyze electrical properties of single coils four different patterns (Z1, W1, Z2, W2) were screen-printed. The thickness and width of the paths were measured using optical profilometer[2] (Fig. 13).

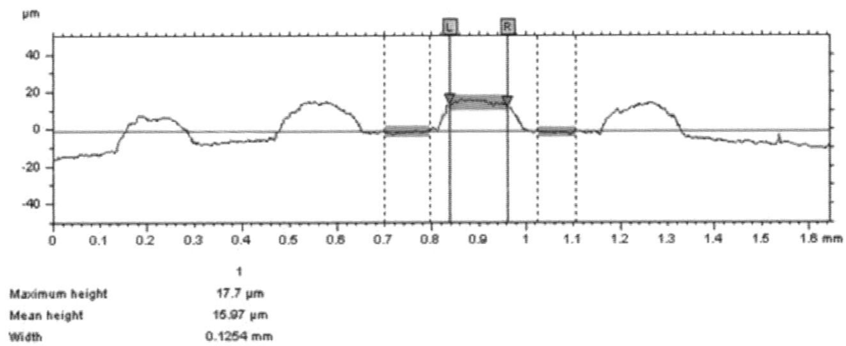

Figure 13. Example profile of W1 coil.

For calculations the average width of paths - 200 µm, average thickness - 16.6 µm and surface resistance 2 mΩ/□ (ESL paste) were taken. Resistance was measured using the Fluke 8842A multimeter. Measured and calculated electrical parameters of the single coils are summarized in Tab. 1.

Table 1. Properties of the single coils.

	Z1	W1	Z2	W2
Length of paths *l* [mm]	46.2	66.0	76.2	80.8
Resistance R [mΩ]	403	476	680	611
Resistivity ρ [Ωm]	$2.90 \cdot 10^{-8}$	$2.39 \cdot 10^{-8}$	$2.96 \cdot 10^{-8}$	$2.51 \cdot 10^{-8}$
Sheet resistance R□ [mΩ/□]	1.75	1.44	1.78	1.51

In order to estimate the inductance of a single spatial coil the Greenhouse [1] (self-inductance) and Hoer [2] (mutual inductance) calculation methods were used. Greenhouse method is dependent on solving inductance equations for planar thin- or thick film coils. Self-inductance must be calculated for every segments of coil. To obtain mutual inductance with Hoer method was need to use a complicated formula which allows determine mutual inductance between rectangular bars. To verify the calculation results inductance of each fabricated coils (ESL paste) was measured using the precision RLC Meter (Agilent E4980A, frequency - 2 MHz). The results are shown in Tab. 2.

Table 2. Measured and calculated properties of the single coils.

	Z1	W1	Z2	W2
Series resistance Rs [mΩ]	430.78	639.34	576.7	666.75
Series inductance Ls [nH]	56.67	88.56	98.34	114.62
Calculated self-inductance [nH]	52.67	70.01	71.87	77.33
Calculated mutual inductance [nH]	3.03	15.56	19.35	31.10
Calculated inductance L [nH]	55.70	85.57	91.22	108.43

[2] Taylor Hobson 3D optical profiler available at the Faculty Division of Microsystems and Photonics, Faculty of Microsystem Electronics and Photonics, Wroclaw University of Technology.

Also the whole structures were measured and calculated. The results are summarized in Tab. 3.

Table 3. Measured and calculated properties of the structures.

	Agilent E4980A (2 MHz)		Fluke 8842A	Resistance R [Ω] calculated for ESL paste (2 mΩ/□)	Resistance R [Ω] calculated for DuPont paste (3 mΩ/□)
	Rs [Ω]	Ls [μH]	R [Ω]		
1.8 D	170.04	63.83	97	-	188.50
1.10 D	249.09	94.02	121	-	235.62
2.8 D	329.13	145.70	127	-	263.76
2.10 D	402.34	171.60	125	-	329.70
1.2 E	31.00	6.10	29	31.42	-
1.8 E	173.38	67.89	113	125.66	-
1.10 E	252.05	100.57	144	157.08	-
1.12 E	352.40	140.28	168	188.50	-
2.2 E	37.51	11.31	34	43.96	-
2.8 E	315.10	161.07	142	175.84	-
2.10 E	586.30	265.67	176	219.80	-
2.12 E	1320.00	420.53	248	263.76	-

5. Conclusions

The self-made system for characterization of electromagnetic microgenerators allows to generate an alternating magnetic field and to measure the magnet disc's rotary speed.

The self-inductance of the single coils was calculated using the Greenhouse method whereas the mutual inductance with the Hoer method. Calculated and measured inductance values are very similar.

The generated power is larger for the microgenerators fabricated using the second pattern (Fig. 2). The best output parameters were achieved for 8-layer structure (second pattern and DuPont paste) – 2.8 D. For the rotary speed of 11827 rpm it generates the power of 589.91 mW (for 167 Ω resistance load).

Acknowledgments

This work was supported by statutory activity of Wrocław University of Technology (S40067 W-12).

References

[1] Greenhouse HM 1974 Design of Planar Rectangular Microelectronic Inductors *IEEE Transactions on Parts, Hybrids and Packaging,* Vol. PHP-10 **2** 101

[2] Hoer C and Love C 1965 Exact Inductance Equations for rectangular Conductors With Applications to More Complicated Geometries *Journal of Research of the National Bureau of Standards – C. Engineering and Instrumentation,* Vol. 69C **2** 127

[3] Gierczak M, Walasek P and Markowski P 2015 Characterization of LTCC-based electromagnetic microgenerators *Elektronika – konstrukcje, technologie, zastosowania,* Vol. 56 **3** 31

[4] Pan C T, Chen Y J 05/2008 Application of low temperature co-fire ceramics on in-plane micro-generator, *Sensors and Actuators A: Physical*, Vol. **144(1)** 144-153

[5] Pan C T, Chen Y J, Shen S 07/2009 Simulation and analysis of electromagnetic in-plane microgenerator, *Journal of Micro/Nanolithography, MEMS and MOEMS*, Vol. **8(3)**

Evaluation of Polyesterimide Nanocomposites Using Methods of Thermal Analysis

B Gornicka[1], L Gorecki[1], K Gryzlo[1], D Kaczmarek[2] and D Wojcieszak[2]
[1]Electrotechnical Institute Division of Electrotechnology and Materials Science, M. Sklodowskiej-Curie 55/61, 50-369 Wroclaw, Poland
[2]Faculty of Microsystems Electronics and Photonics, Wroclaw University of Technology, Janiszewskiego 11/17, 50-372 Wroclaw, Poland

E-mail: gornicka@iel.wroc.pl, damian.wojcieszak@pwr.wroc.pl

Abstract. Polyesterimide resin applied for winding impregnation has been modified by incorporating the hydrophilic and hydrophobic nanosilica, montmorillonite and aluminium oxide. For assessment of the resins in liquid and cured states thermoanalytical methods TG/DSC were used. For pure and nanofilled resins the results of investigation of AFM topography, bond strength, dielectric strength and partial discharge resistance have been also presented. It was found that dielectric and mechanical properties of polyesterimide resin containing hydrophilic silica as well aluminium oxide were much improved as compared to pure resin. Based on our investigations we have found that the methods of thermal analysis may be very useful for evaluation of nanocomposites: DSC/TGA study of resins in the liquid state under dynamic conditions can be applied to pre-select nanocomposites; isothermal TG curves of cured resins can be utilized for thermal stability evaluation; in turn, TG study after thermal ageing of cured resins could confirm the barrier properties of nanocomposites.

1. Introduction

It is well known that polymer nanocomposites exhibit improved properties relative to unfilled polymers. For many nanocomposites have been noted an increase in thermal stability [1-6] and better mechanical [6] and dielectric properties [7-8]. One of the first developed nanodielectrics was "corona-resistant" enamel for winding wires [9 -11]. In turn, the article presents the results of investigation of polyesterimide impregnating resin that was filled using various nanoparticles [12]. The aim of our studies was to show the possibilities of using thermal analysis methods for the assessment of nanocomposites. The mechanical and dielectric properties of nanofilled resins as well as surface topography were also presented.

2. Experimental

Four kinds of nanofillers, i.e.: aluminium oxide Alu C, hydrophobic nanosilica Aerosil R 202, hydrophilic nanosilica Aerosil 200 (all from Evonik) and montmorillonite MMT Rheospan AS (Nanocor) in an amount of 1.5wt% have been added to base polyesterimide resin (Polifarb CW) and was dispersed using a method described in our patent [13].

For all the pure and nanofilled resins it was performed: (i) the TG/DTG/DSC study for resins in a liquid state (temperature range 23 up to 800°C, a heating rate of 10K/min, air atmosphere) using a thermogravimetric analyser TGA/DSC1 (Mettler Toledo); (ii) the bond strength test acc. to IEC

60455-2; (iii) dielectric strength test acc. to IEC 60455-2; (iv) determining of partial discharge resistance (lifetime) under pulse voltage 1100V/20 kHz. Moreover, for two kind of cured PEI samples, both the pure and nanofilled with hydrophilic nanosilica resins the following results of investigation were presented: (i) surface topography by the atomic force microscopy (AFM) with a scanning size of 1 μm. Tested surface was prepared by breaking in order to obtain natural fracture; (ii) the TG curves under isothermal conditions, i.e. at 240°C, 260°C and 280°C; (iii) bond strength after various periods of thermal ageing at 250°C; (iv) the TG curves under dynamic conditions after thermal ageing at 210°C.

3. Results and discussion

The results of the investigation of bond strength, dielectric strength and partial discharge resistance under pulse voltage 1100V/20 kHz testing of pure and nanofilled resins are compared in Table 1.

Table 1. The results of testing of mechanical and dielectric properties of pure and nanofilled resins.

Properties	Pure PEI resin	PEI resin + MMT	PEI resin + hydrophobic nanosilica	PEI resin + hydrophilic nanosilica	PEI resin + aluminium oxide
Bond strength at 23°C*) [N]	184	169	144	230	205
Dielectric strength at 180°C**) [kV/mm]	40.2	33.8	19.0	72.8	62.3
Partial discharge resistance (lifetime) under pulse voltage 1100V/20 kHz**) [min]	45	41	23	254	479

*) the average of 10 measurements
**) the average of 5 measurements

It may be noted that the resins containing hydrophilic nanosilica and aluminum oxide show a marked improvement of the bond strength, dielectric strength and partial discharge resistance whereas the resins containing montmorillonite and hydrophobic nanosilica have the properties much worse than pure PEI resin. The TG/DTG/DSC thermograms of these resins in liquid state are presented in Fig. 1. The resins with hydrophilic nanosilica and aluminium oxide have showed unchanged the TG and DSC curves as compared to pure resin. On the other hand these resins whose properties deteriorated (i.e. with MMT and with hydrophobic nanosilica) show a clear distortion of the process of solvent evaporation and gelling up to 150°C (Fig. 1); the rate of solvent loss was reduced and therefore copolymerization process has been disturbed. So it can be concluded that the thermoanalytical studies of resins modified with nanofillers in the liquid state can be useful in pre-selection of nanofilled resins.

The AFM topography images of samples of pure and nanofilled with hydrophilic nanosilica PEI resins are compared in Fig. 2 and 3, respectively (scanning size: 1 μm). For pure sample, obtained results revealed usually the regular step-like structures at submicrometer level (Fig. 2). The topography of nanofilled sample is quite different; it shows densely packed array of similar objects (Fig. 3) with the grain size of about 10-30 nm. A weaker tendency of creating regular structures during cracking of nanocomposite could suggest better mechanical properties.

For two PEI resins in cured state: pure and modified with hydrophilic nanosilica the TG curves under isothermal conditions, i.e. at 240°C, 260°C and 280°C during period of 10 h, are presented on Fig. 4. It can be seen that mass loss of the nanofilled resin at all test temperatures is slower. It can prove that the thermal stability of nanofilled resin is better. In nanocomposite, due to the barrier effect of the nanofiller particles, on the one hand a mass transfer of degrading polymer to the vapour phase and on the other hand, oxygen transfer through polymer could be inhibited [14].

IOP Conf. Series: Materials Science and Engineering **113** (2016) 012012 doi:10.1088/1757-899X/113/1/012012

The improved thermal stability of the PEI resin modified with nanosilica was confirmed by investigation of the bond strength values after various periods during thermal ageing at 250°C. After aging period of 1700 h the nanofilled resin showed the clearly higher level of bond strength (23.1 N) than pure one (11.2 N).

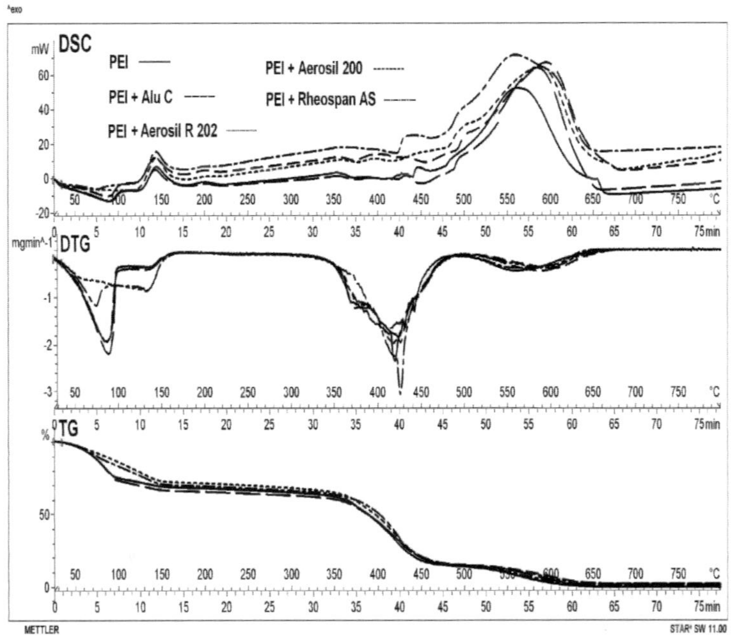

Figure 1. The TG/DTG/DSC curves of the pure and modified with Al_2O_3, hydrophobic and hydrophilic SiO_2, and MMT resins in liquid state

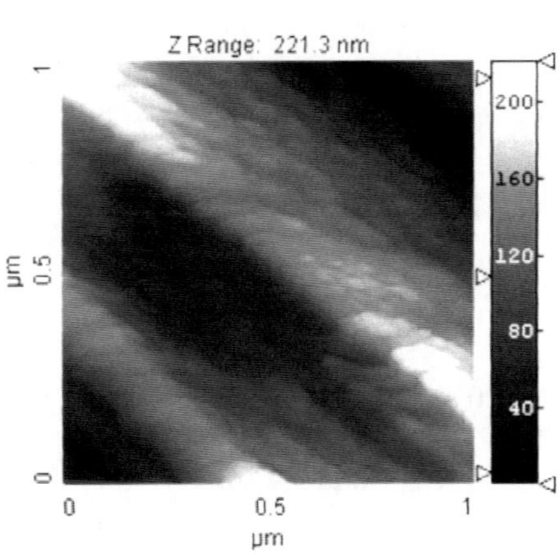

Figure 2. AFM image of the pure PEI resin.

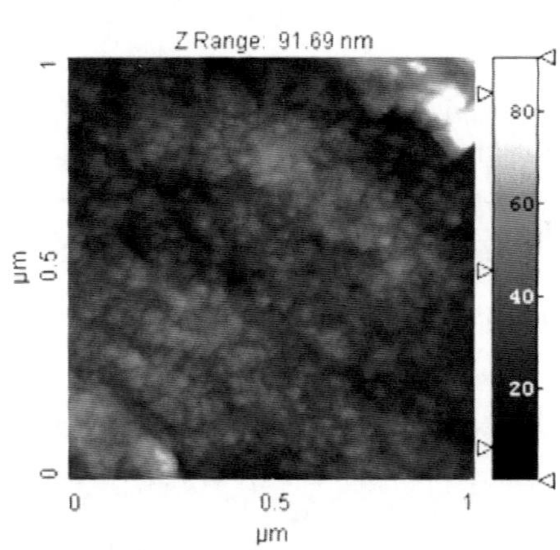

Figure 3. AFM image of the PEI resin with hydrophilic nanosilica.

The TG curves of both pure and filled with hydrophilic nanosilica PEI resins have been determined after various ageing periods at raised temperatures. It was found that very high abnormal peaks appear always on the TG curves of the pure resin at about 420°C. As example, the TG curves of the resins after period 840 h of thermal ageing at 210°C are shown in Fig. 5. The cause of the occurrence of these peaks may be a violent releasing of gaseous products of thermal ageing. However this phenomenon does not appear in the case of nanofilled resin, probably because of the barrier properties of the nanocomposite (Fig. 5).

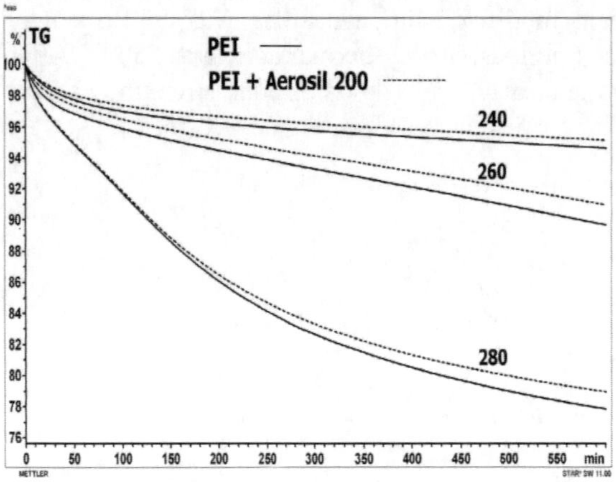

Figure 4. The TG thermograms of both pure and modified with hydrophilic nanosilica PEI resins under isothermal condition.

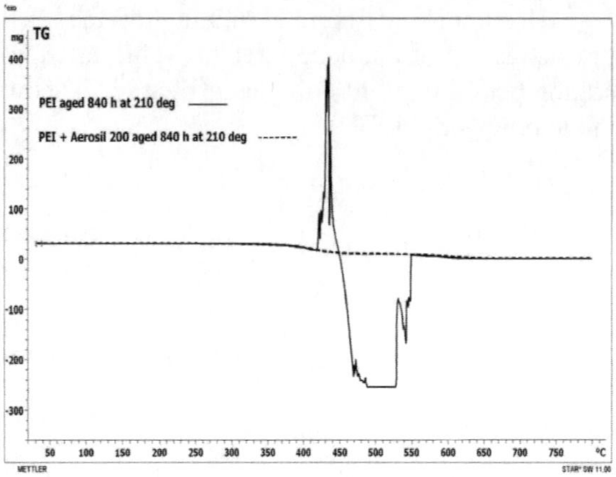

Figure 5. The TG thermograms of both pure and modified with hydrophilic nanosilica PEI resins after 840 h ageing at 210°C

4. Conclusions

Polyesterimide resin filled with nanosilica, montmorillonite and aluminium oxide have been investigated. The significant improvement of dielectric and mechanical properties of resins comprising a hydrophilic nanosilica and aluminium oxide was obtained. It was found also that the thermal analysis methods can be very useful in assessment of nanocomposites, i.e.: (i) the TG/DTG/DSC study of resins in the liquid state can be used to preliminary assess nanofilled resins; (ii) TG study of cured resins at isothermal conditions can be applied to evaluate the comparative thermal stability of developed nanocomposites; (iii) TG study of pure and nanofilled PEI resins after thermal ageing may be the evidence of the barrier properties of nanocomposites.

5. References

[1] Golebiewski J, Galeski A, 2007 *Compos. Sci. Technol* **67** 3442-7
[2] Lewicki J P J, Liggat J, Pethrick R A, Patel M., Rhoney I, 2008 *Polym. Degrad. Stab.* **93** 158-68
[3] Kashiwagi T, Morgan A B, Antonucci J M at al. 2003 *J. Appl. Polym. Sci.* **89** no 8:2072-8
[4] Chrissafis K, Paraskevopoulos K M, Pavlidou E, Bikiaris D, 2009 *Thermochim. Acta* **485**:65-71
[5] Leszczyńska A, Njuguna J, Pielichowski K, Banerjee J R, 2007 *Thermochim. Acta* **453**:75-96
[6] Gornicka B, Prociow E, 2009 *Acta Phys. Pol.* **115** no 4:842-5
[7] Tanaka T, 2005 *IEEE T. Dielect. El. In.* 12 no 5:914-28
[8] Nelson J K, 2007 *El. In. Conf. El. Manuf. Expo* **22**-24:229-35
[9] Gornicka B, Prociow *K* 2009 *Mat. and Manuf. Proc.* **24** 10/11 1202-1206
[10] Zhang G, Wu G, Tong L, He E 2005 *Proc. of 2005 Symp. Elec. Insul. Mat.* 180-183
[11] Bohm F R, Nagel K, Schindler H 2003 *El. Insul. Conf. and El. Manuf. & Coil Winding Techn. Conf., Proc.* 109-113
[12] Winkeler M, Goetter R W 2009 IEEE EIC 401-404
[13] Gornicka B, Czolowska B, Gorecki L, *Patent PL 214577 B1*
[14] Beyler C L, Hirshler M M, http://gbhint.tripod.com/papers_5_13_02/368_Beyler_Hirschler_SFPE_Handbook_3.pdf

Acknowledgments
This work was supported by the WRC EIT+ NanoMat under grant "The Application of Nanotechnology in Advanced Materials" co-financed by the EU from the resources of EFRD.

IOP Conf. Series: Materials Science and Engineering **113** (2016) 012013 doi:10.1088/1757-899X/113/1/012013

Study of electrokinetic properties of nanofillers used for the preparation of nanocomposites

B Górnicka, K Gryzło, L Górecki

Electrotechnical Institute Division of Electrotechnology and Materials Science, M. Sklodowskiej-Curie 55/61, 50-369 Wroclaw, Poland

E-mail: gornicka@iel.wroc.pl

Abstract. The aim of the studies was to determine, how the zeta potential and particle size of nanofillers affect the stability of the dispersion of nanoparticles in the electroinsulating varnish, and consequently can influence the varnish properties. We have investigated the nanofillers that had been used for the preparation of nanocomposites in our earlier works. It has been found that in toluene the zeta potential of all nanofillers is much increased. It enables modifying of nanofillers. Correlation between the properties of the nanocomposite and the results of the particle size measurements has been observed. Thus studies of electrokinetic properties can be used during the preparation of nanocomposites, as well to determine the factors affecting the nanoparticles deagglomeration and a preliminary determination of the optimal technological parameters of process of preparing of nanocomposites.

1. Introduction

Polymer nanocomposite generally comprises very small filler particles dispersed in the polymer matrix. If the nanocomposite is prepared by direct mixing of the filler into a polymeric matrix in solution or in melt, we can get a colloidal dispersion. A key indicator of the stability of colloidal dispersions is the zeta potential. The magnitude of the zeta potential indicates the degree of electrostatic repulsion between adjacent, similarly charged particles in dispersion. Colloids with high zeta potential (negative or positive) are electrically stabilized while colloids with low zeta potentials tend to coagulate or flocculate. Zeta potential measurements can contribute to improve formulation stability and reduce formulation time and cost [1-7].

We have studied the zeta potential and particle size of nanofillers that were used for the preparation of nanocomposites in our earlier works. The aim of our research was to determine in what way the electrokinetic properties of nanofillers affect the stability of the dispersion of nanoparticles in the electroinsulating polyesterimide and influence the varnish properties.

2. Experimental

Zeta potential and particle size of various nanofillers in different solvents was measured using Zetasizer Nano ZS Malvern Instruments. The following commercial nanoparticles have been studied: (i) fumed nanosilicas: Aerosil 200 (hydrophilic), Aerosil 380 (hydrophilic), Aerosil R202 (hydrophobic, after treated with a polydimethylsiloxane) – the all from Evonik Industries AG as well as Orisil 200 (hydrophilic, Orisil-Farm, Ukraine); (ii) nanopowders of metal oxides: Aeroxide Alu130, Aeroxide TiO_2P90 and VPZnO20 (Evonik Industries AG); (iii) montmorillonites: Rheospan AS31740 and Nanocor I.22E (Nanocor); (iv) conductive nanofillers: carbon nanotubes High Purity MWNT and

Industrial Grade MWNT as well as a nanosilver 47MN-03 (Inframat Advanced Materials). The zeta potential study was performed as a function of pH of water. Nanoparticles dispersed in polar solvents (water, methanol) and non-polar solvent (toluene) were also tested. Furthermore, we have determined the ζ-potential as a function of time in solvents. Then we have attempted to relate the obtained electrokinetic parameters to properties of nanocomposites based on polyesterimide varnish with mentioned nanoparticles.

3. Results and discussion

The influence of water pH on the zeta potential of various nanoparticles is presented in Fig. 1-4. Curves of zeta potential versus pH of hydrophilic nanosilicas are similar, whereas the curve of hydrophobic Aerosil R 202 shows two isoelectric points (Fig. 1). Metal oxide particles exhibit large changes in the zeta potential with the change of the pH values of the solution, going from positive values to negative (Fig. 2). Zeta potential is negative in the montmorillonites, only for Nanocor I.22 at lowest pH value is positive (Fig. 3). The same is true for conductive nanoparticles (Fig. 4).

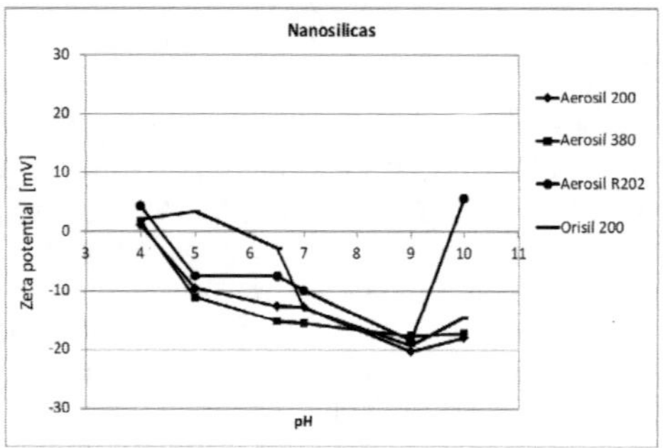

Figure 1. Zeta potential versus pH curves for nanosilicas

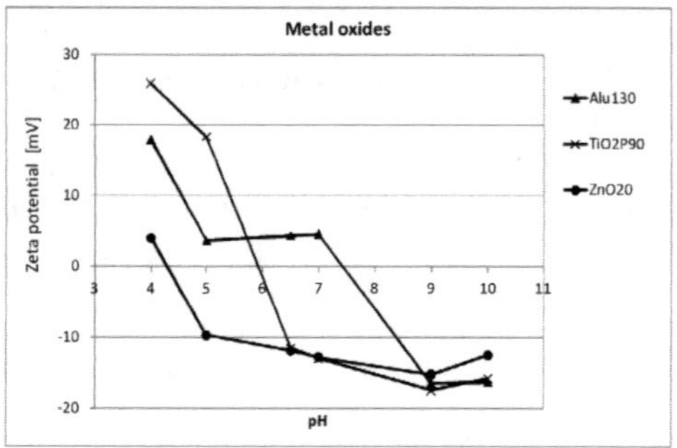

Figure 2. Zeta potential versus pH curves for metal oxides nanoparticles

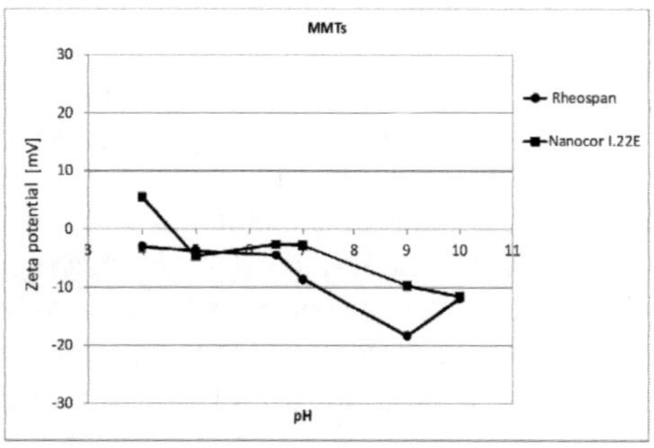

Figure 3. Zeta potential versus pH curves for montmorillonites

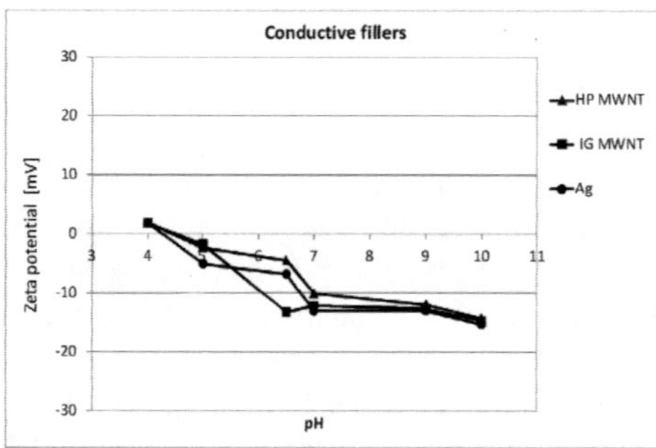

Figure 4. Zeta potential versus pH curves for conductive nanoparticles

The maximum of absolute value of zeta potentials of the nanofillers are presented in Fig. 5. The value of ζ-potential of the all nanoparticles was less than 30 mV, so they did not form a stable colloid in water.

Absolute zeta potential of the nanoparticles dispersions in various solvents including pure water, methanol and toluene is illustrated in Fig. 6. It is clear that ζ-potential of the all nanofillers much increases in toluene. This increase is particularly high for carbon nanotubes, montmorillonites and aluminum oxide, so that toluene can be used to modifying of nanofillers [7].

In turn, measurements of the nanofillers particle size in water confirmed the formation of agglomerates. For example, nominal size of grain of nanosilicas ranges from a few to several micrometers while the measured size is more than one orders greater (Fig. 7).

It was observed that residence time of nanofiller in a solvent influences the average particle size of hydrophilic and hydrophobic nanosilica. It was shown in Figures 8 and 9 for hydrophilic and hydrophobic nanosilica.

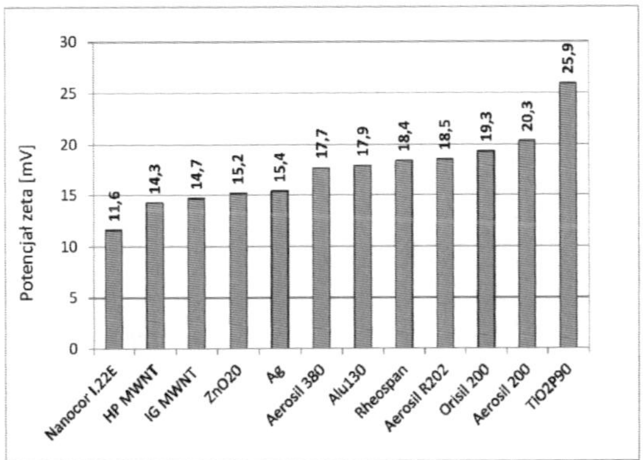

Figure 5. Absolute value of maximum ζ-potential in water for various pH values

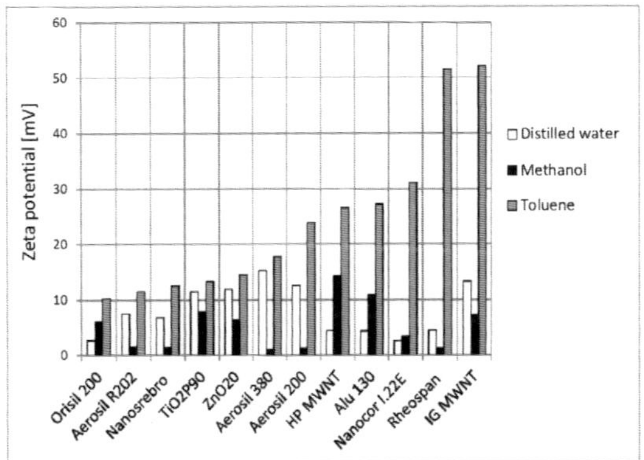

Figure 6. Absolute ζ-potential in various solvents

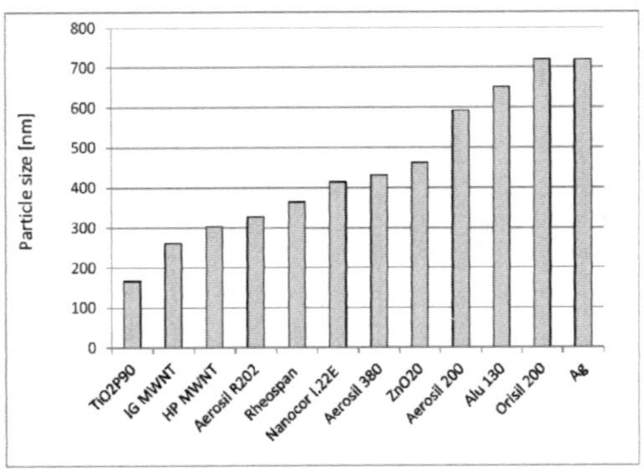

Figure 7. Measured Average particle size of nanofillers

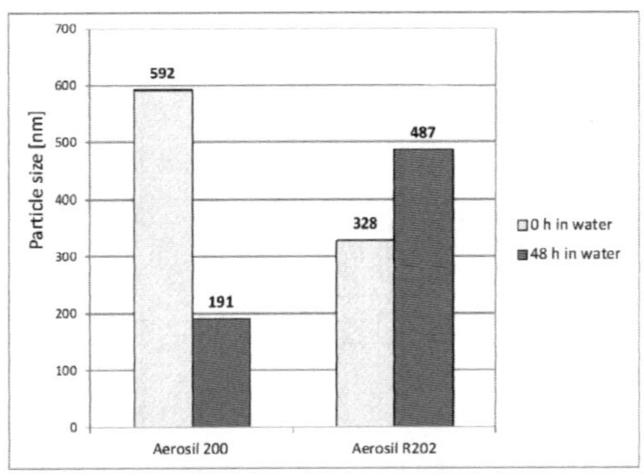

Figure 8. Average particle size of the silica nanoparticles in water, and after 48 h in water

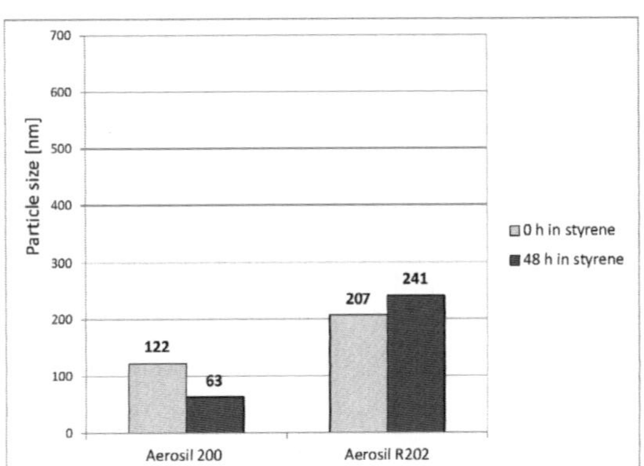

Figure 9. Average particle size of the silica nanoparticles in styrene, and after 48 h in styrene

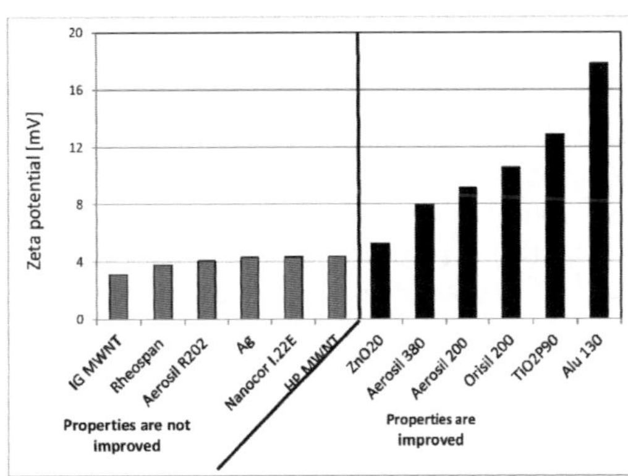

Figure 10. Absolute zeta potential of the nanoparticles in water after 48 hours

The average particle size of hydrophilic nanosilica (Aerosil 200) is distinctly diminished (deagglomeration) after period of 48 h staying in a solvent such as water or styrene. This phenomenon does not occur for the hydrophobic nanosilica (Aerosil R 202). A beneficial effect on the nanocomposite seems to have using of a stabilization period of certain nanofillers in the varnish.

Polyesterimide varnishes filled with the nanoparticles had been investigated in our previous works as it has been mentioned above. A qualitative assessment of the properties of varnishes filled by the nonconductive nanoparticles compared with the properties of the base varnish is shown descriptively in Table 1. The table comprises also the values of absolute zeta potential of nanofillers in water and in water after period of 48 hours. Correlation between zeta potential nanofillers in the water after 48 h and the properties of nanocomposites contained the tested nanofillers has been found.

Absolute zeta potential of the nanoparticles dispersions in water after 48 hours is presented in Fig. 10. On the right side of a black line are nanoparticles which, when introduced using the same method of dispersing into the varnish have resulted in improved properties of the nanocomposites. This could mean that by examining the zeta potential in water after 48 hours, we can preselect the nanofillers.

Table 1. Summary of the properties of nanofilled varnishes that contain various nonconductive nanoparticles and their zeta potential in water as well after period of 48 hours in water.

Nanofiller	Mechanical properties	Dielectric properties		Gelling process (TGA/DSC)	Absolute ζ-potential in water [mV]	Absolute ζ-potential in water after 48 h [mV]
	Bond strength	Dielectric strength	Partial discharges resistance			
Rheospan AS	+	–	–	disturbed	8,3	3,8
Aerosil R202	0	–	–	disturbed	7,5	4,1
NanocorI.22E	0	0	–	disturbed	1,7	4,3
VP ZnO20	++	0	++	normal	9,8	5,3
Aerosil 380	+	+	+++	normal	6,8	7,9
Aerosil 200	++	++	+++	normal	12,6	9,1
Orisil 200	++	++	++	normal	7,1	10,6
TiO_2 P90	+	++	+	normal	9,6	12,9
Alu 130	++	++	+++	normal	20,7	17,8

– worse; – – significant worse; 0 no changes;
+ improved; ++ much improved; +++ very much improved.

4. Conclusions

Study of influence of electrokinetic properties of nanofillers used for preparing of nanofilled varnishes indicate that: (i) None of used nanofillers does not form a stable colloid in water at different pH and a correlation between zeta potential nanofillers in the water with properties of nanocomposites was not found; (ii) the particle size of all nanofillers measured in the water indicate the formation of agglomerates; (iv) For all nanofillers ζ-potential much increases in toluene; (v) correlation can be found between the particle size and the properties of nanocomposites; (vi) A clear correlation between the potential of nanofillers after 48 in water with the properties of the resulting nanocomposites has been found.

5. References

[1] Delgado A V, González-Caballero F, Hunter R J, Koopal L K, Lyklema J 2007 *J. Colloid Interf. Sci.* **309** 2:194–224
[2] Colic M, Fuerstenau D W 1998 *Powder Technol.* **97** 129-138
[3] Vallar S, Houivet D, El Fallah J, Kervadec D, Haussonne J-M 1998 *J. Eur. Ceram. Soc.* **19** 1017-1021
[4] [4] Tantra R, Schulze P, Quincey P, 2010 *Particuology* 8 79–285
[5] Butterworth M D, Corradi R, Johal J, Lascelles S F, Maeda S, Armes S P 1995 *J. Colloid Interf. Sci.* **174** 2:510–517
[6] Zielecka M, 2011 *Polimery* **56** 10 765-768
[7] Lewis S, Haynes V, Wheeler-Jones R, Sly J, Perks R M , Piccirillo L 2009 *Thin Solid Films* 1-5

The research of ceramic materials for applications in the glass industry including microwave heating techniques

K.Kogut, K.Kasprzyk, B.Zboromirska-Wnukiewicz, T.Ruziewicz,
Electrotechnical Institute Division of Electrotechnology and Materials Science Wrocław,
Skłodowskiej – Curie Street 55/61, 50-369 Wrocław, Poland

E-mail: k.kogut@iel.wroc.pl

Abstract. The melting of a glass is a very energy-intensive process. Selection of energy sources, the heating technique and the method of heating recovery are a fundamental issue from the furnace design point of view of and economic effectiveness of the process. In these processes the problem constitutes the lack of the appropriate ceramic materials that would meet the requirements. In this work the standard ceramic materials were examined and verified. The possibilities of application of microwave techniques were evaluated. In addition the requirements regarding the parameters of new ceramic materials applied for microwave technologies were determined.

1. Introduction

The microwave sintering initiated in mid 60th is currently an area of extensive research [1]. The microwave technique enable the fast and uniform heating of small and big shapes, removing volatile elements (binders, moisture, etc.) from big volumes and allow to reduce the thermal stresses in the heated material.

Rising interest in the melting and sintering process via microwave is a results of time saving and a consequence of more homogeneous structure and improved properties of final products. The microwave heating of materials is completely different from the conventional heating method, which the energy transfer to inside of the material is a result of thermal conductivity [2].

Microwave processing of the materials, includes heating and sintering, is fundamentally different from the conventional processing involving radiant/ resistance and/ or convention heating followed by transfer of thermal energy via conduction to the inside of the work – piece through thermal conductivity mechanism.

Figure 1 illustrates some distinguishing features between the conventional and the microwave heating.

Great advantage of the microwave technologies application in relation to conventional methods, during glass melting, are significantly reduced emission levels of harmful substances to the atmosphere by elimination or at least mitigation of mine raw materials used. The decrease of the emission of greenhouse gases emission is a necessary action for slowing down the global climate change. In the case of furnaces supplied by fossil fuels, the realized thermal energy is connected with fuel combustion but also with emission of CO_2, SO_2, NO_x and dusts.

In figure 2 types of influences of microwaves on different kinds of materials are presented.

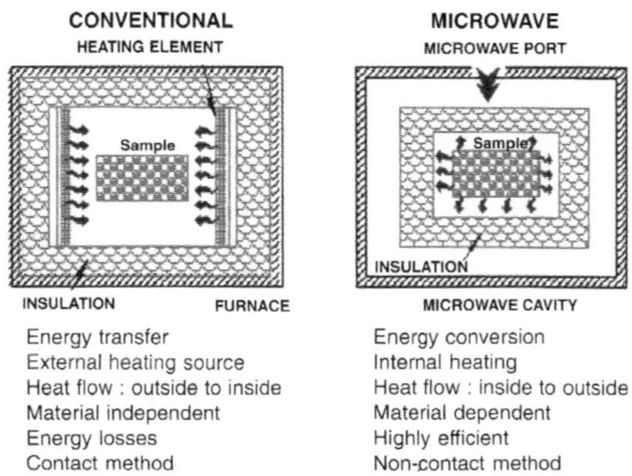

Figure 1. Comparison of heating procedure between conventional and microwave methods [3].

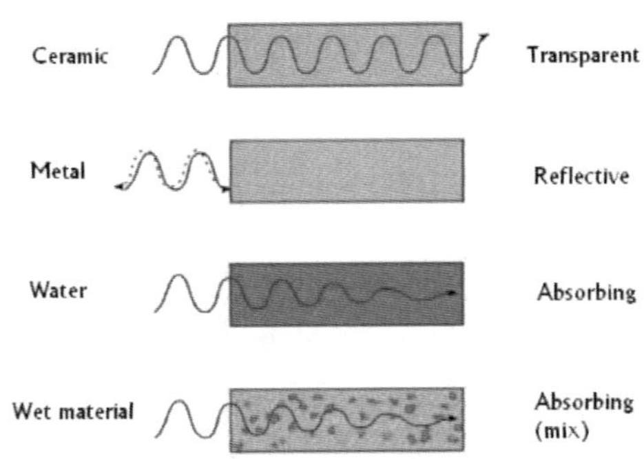

Figure 2. The results of microwaves radiation for various kinds of materials [4].

Figure 3. The example of microwave device (microwaves generator) used for melting of the glass. Corundum crucible.

Significant barrier for applying the microwave technology in the process of glass melting is the lack of the appropriate ceramic materials meet requirements:
- resistance against temperatures up to 1600 ° C,
- stress resistance temperature (desirable low thermal expandability),
- wearability through the stream of the molten glass,
- corrosion resistance in gasses emitted by the molten glass,

- appropriate properties in the microwave field – zero penetrability microwaves, or optionally full transparent for microwaves in the highest temperatures.

2. Materials and methods

In this work the available ceramic materials were examined. Modified materials for application in microwave heating techniques were prepared. These materials were heat-treated. Based on the literature [5, 6], the adequate temperature of sintering was applied. A density and a porosity of the samples were determined by hydrostatic weighing method.
For research purposes the materials based on the alumina oxide, magnesia and the soapstone characterized by a low dissipation dielectric factor talking about the possibilities of heating materials in high frequencies, were used (tab. 1).

Table 1. Various materials used during experiments

	Material A, A_L	Material B, B_L	Material C, C_L	Material D, D_L	Material E - Steatite
Composition based on:	• MgO • Al_2O_3 – Martinswerk • Clay "Jaroszow"	• Al_2O_3 – Hungarian • Clay "Jaroszow"	• Al_2O_3 – Martinswerk • Clay "Jaroszow"	• MgO • Al_2O_3 – Hungarian • Clay "Jaroszow"	• Talc, • Plastic Clay • Feldspar

A_L, B_L, C_L, D_L – preliminary cooled to – 23 °C,

Dissipation dielectric factor value was measured with aid of RF impedance/ Material analyzer Agilent E4991A.
A microwave device with the power of 1 kW was used to measure the speed of ceramic materials heating. Samples in the form of popular ceramic fittings were under constant microwave influence by the period of 30 minutes. The increase of the temperature by the pyrometer Abatronic AB infrared – 8859, with measurement range of temperatures from -50 °C to 1600 °C and the 50:1 optical resolution was carried out.
Samples were also tested to evaluate mechanical strength. This particular property is important from the ceramic materials point of view e.g. as high-temperature crucible pots to microwave furnaces. The mechanical strength was appointed by three point bending tests with the use of the Rauenstein machine type 2143 with the measuring scope from 0 - 5 kN. Samples in the form of flattened circular cross – section bars with dimensions about 9.0 x 10.0 x 75 mm were used for examinations.

3. Results

3.1. Bulk density, porosity and absorbability

Bulk density, porosity and absorbability were determined with the use of the Polish Standard *PN-89/E-06307* [7]. The results are showed in figure 4.

The highest density due to the lowest porosity and the absorbability was measured for sample B, sintered at temperature of 1430°C. In order to obtain high porosity [9, 10] to limit the heating of materials every type of samples was cooled to the temperature -23°C. The highest porosity was measured for the sample B_L sintered at temperature of 1100°C.

IOP Conf. Series: Materials Science and Engineering **113** (2016) 012014 doi:10.1088/1757-899X/113/1/012014

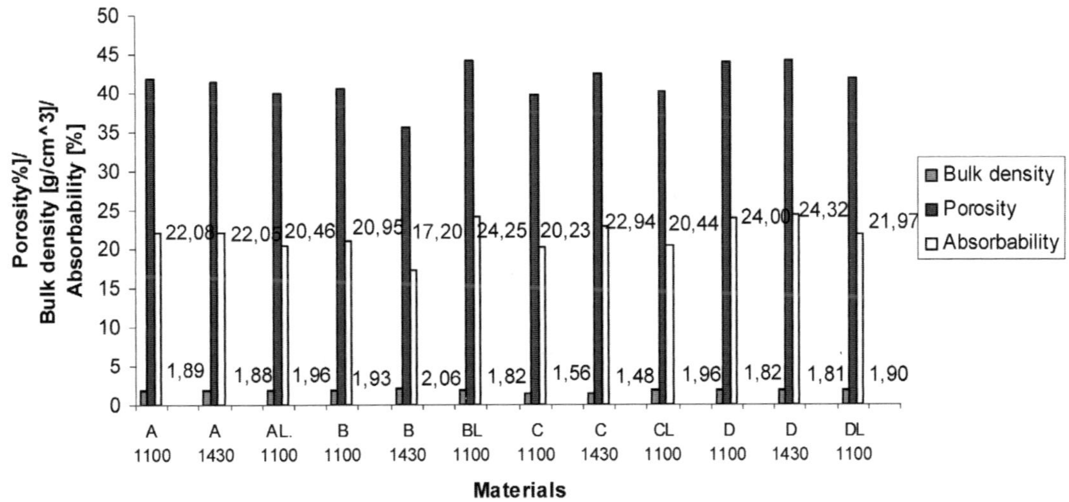

Figure 4. The results of the density, porosity and absorbability of the used materials.

3.2. The electric permittivity and tg delta measurements

Results of the electric permittivity and tg delta of measurements for particular materials are showed in figure 5 and in table 2.

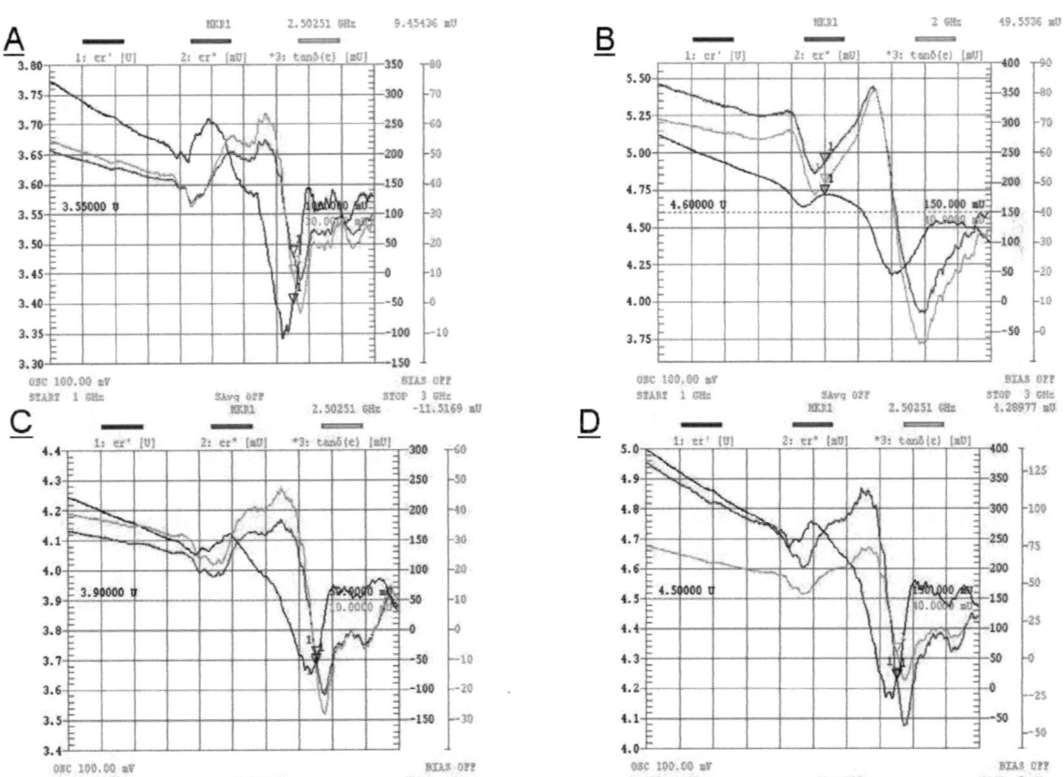

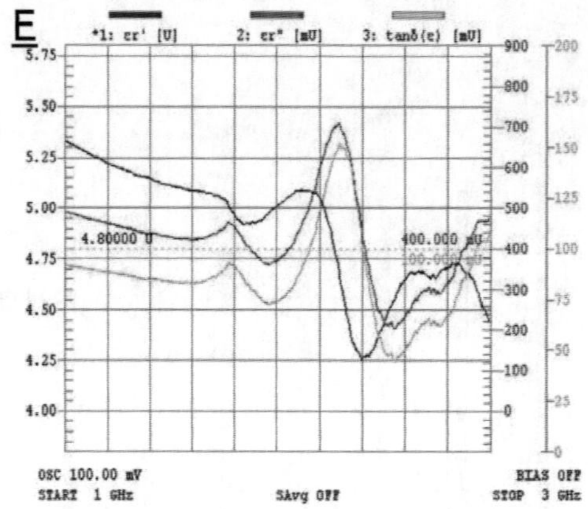

Figure 5. Examples results of the electric permittivity and tg delta of measurements for materials A, B, C, D, E.

Table 2. Results of the electric permittivity and tg delta of measurements in high frequencies for examined materials

		Frequency [GHz]			
		1,00	2,00	2,50	3,00
Material A	ε'	3,83	3,68	3,44	3,59
sintered in	ε'' x10^{-1}	3,29	2,03	1,03	1,44
1100 °C	tgδ x10^{-3}	86,1	55,2	29,8	40,2
Material B	ε'	5,12	4,72	4,26	4,38
sintered in	ε'' x10^{-1}	3,67	1,12	4,38	1,44
1100 °C	tgδ x10^{-3}	71,6	49,3	9,74	33,0
Material C	ε'	4,37	4,18	4,09	4,11
sintered in	ε'' x10^{-1}	2,83	1,88	1,00	1,57
1100 °C	tgδ x10^{-3}	64,7	45,0	24,4	38,1
Material D	ε'	5,00	4,75	4,23	4,46
sintered in	ε'' x10^{-1}	3,80	2,33	2,73	1,17
1100 °C	tgδ x10^{-3}	75,9	49,0	6,46	26,3
Material E	ε'	5,33	5,01	4,42	4,42
sintered in	ε'' x10^{-1}	4,91	3,70	2,29	4,74
1280 °C	tgδ x10^{-3}	92,0	74,0	51,8	107

For comparison, in table 3 the results of the electric permittivity and penetrability for popular ceramic materials applied in microwave heating techniques, were presented.

Table 3. Influence of microwaves on the popular ceramic materials [4].

Type of the materials	ε'	ε"	Temperature [°C]	Penetration depth of microwaves [cm]
Alumina Oxide (Al_2O_3)	9,00	0,004	Room	1460
Alumina Oxide (Al_2O_3)	9,46	0,01	296	600
Alumina Oxide (Al_2O_3)	10,15	0,055	683	113
Alumina Oxide (Al_2O_3)	11,18	0,241	1221	27
Silica glass (SiO_2)	3,78	0,00023	25	16700
Sapphire (Al_2O_3)	9,40	0,00023	Room	26474
Silicon carbide (SiO_2)	3,50	0,00035	Room	10410
Silicon carbide (SiO_2)	4,00	0,04	2000	97
Zirconium Oxide	18,00	2,34	300	4
Zirconium Oxide	22,30	8,25	800	1

3.3. Investigation of the speed of heating for the commonly used ceramic materials

A microwave device with the power of 1 kW was used to investigate the speed of heating for the commonly used ceramic materials and materials prepared in this work (tab.1). Samples were put into constant microwave influence by the period of 30 minutes. Results of measurements are showed in figure 6 and 7.

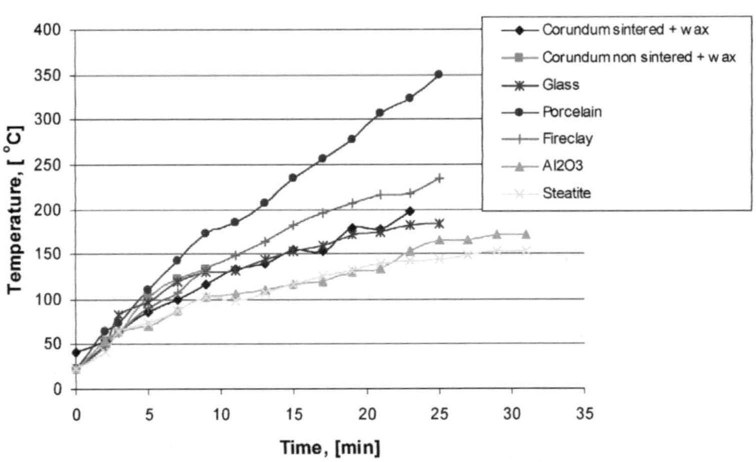

Figure 6. Speed of the common ceramic materials heating under the influence of microwaves.

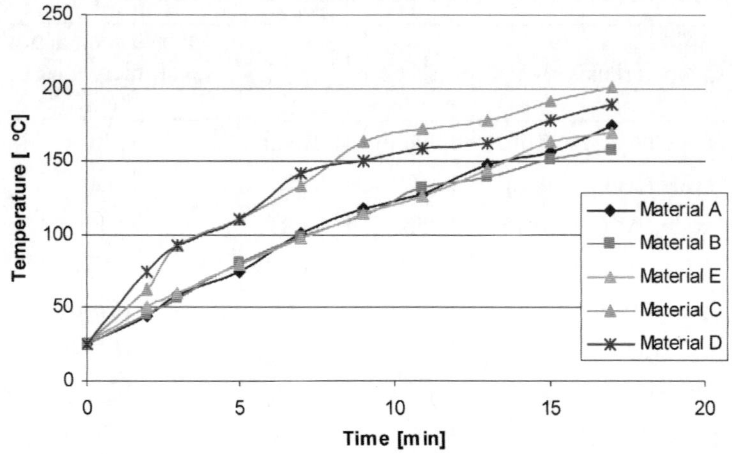

Figure 7. Speed of the ceramic materials A, B, C, D, E heating under the influence of microwaves.

It was found that materials prepared in the work based on the Al_2O_3, MgO and steatite are characterized by lower temperature of heating in the same time and at the same power of the microwave device, than other popular ceramic materials. The samples at high frequencies have a low dielectric loss factor and high porosity. Therefore, the heat generated in material by the dielectric loss factor is lower than the other samples.

3.4. The mechanical bending strength

The mechanical bending strength was determined for the 5 samples, each of the measured materials. To calculate the mechanical bending strength the following equation according to PN IEC 60672 – 2 clause 6.5 [8], was used:

$$Rg = \frac{8 \cdot F_s \cdot l}{\pi \cdot s \cdot d^2} [MPa] \qquad (1)$$

where:
Fs – bending force,
s, d, l – dimensions of the bars respectively: width, height, length

The results are shown in figure 8.

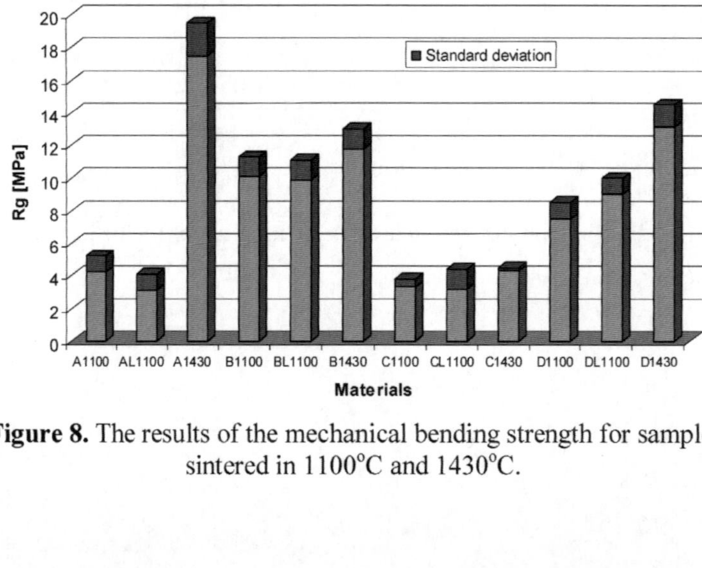

Figure 8. The results of the mechanical bending strength for samples sintered in 1100°C and 1430°C.

It was noticed the increasing trend of higher mechanical bending strength for samples sintered at temperature 1430 °C then samples sintered in 1100 °C. The largest impact on the value of the mechanical strength was observed for the sample of material A. The addition of magnesium oxide and high quality alumina oxide with fragmentation 80% <10 μm, along with sintering at the temperature above 1400 °C improves the Rg value several times.

4. Conclusions

- The microwave techniques intended for sintering or melting of the glass can save the energy spent during the heating process. In addition the advantage is lower temperature or shorter time of sintering.
- Ceramic materials based on the alumina oxide, magnesia and the steatite during attempts with using the microwaves generator with the power of 1 kW are heated up to lower temperatures than other examined materials. The samples at high frequencies have a low dielectric loss factor and big porosity.
- Materials with lower rate of dissipation factor weren't heated up so intensively like another's. These losses are connected with dispersion of the energy via warming of the dielectric.
- Samples sintered in the same temperatures, preliminary cooled to the temperature - 23 °C were characterized by higher porosity than non cooled samples,
- Sintering of the samples based on magnesium oxide, high quality alumina oxide with fragmentation 80% <10 μm, at temperature above 1400 °C, improves the quality of this process and mechanical bending strength of this materials.

5. References

[1] Menezes R, Souto P, Kiminami R 2012 *Sintering of ceramics – New Emerging Techniques. Microwave Fast sintering of ceramic materials* chapter 1 3–26
[2] Oghbaei M, Mirzaee O 2010 *J. Journal of Alloys and Compounds* **494** 175-189
[3] Agrawal D 2006 *J. Transactions Of The Indian Ceramic Society* **65 (3)** 129-144
[4] http://wichary.eu
[5] Fu P, Lu W, Lei W, Xu Y, Wang X, Wu J 2013 Transparent polycrystalline $MgAl_2O_4$ ceramic fabricated by spark plasma sintering: Microwave dielectric and optical properties *J. Ceramics International* **39** 2481-2487
[6] Zhou D, Xueman P, Kai C, Ziliang W 2012 Enhanced dielectric properties of alumina ceramic substrate for microwave application *J. IEEE International Symposium on Radio-Frequency Integration Technology (RFIT)* 101-103
[7] Polish Standard *PN-89/E-06307 Electroinsulation ceramic materials*
[8] Polish Standard PN IEC 60672 – 2 Insulation materials: ceramic materials and glass
[9] Gregorová E, Past W, 2011, *Journal of the European Ceramic Society* **31**, 2073–2081
[10] Xue W, Sun Y, Huang Y, Xie Z, Sun J, Preparation and Properties of Porous Alumina with Highly Ordered and Unidirectional Oriented Pores by a Self-Organization Process, 2011 *J. Am. Ceram. Soc.*, **94 [7]** 1978–1981

Electrical properties of varistors based on tin oxide.

W Mielcarek[1], **K Prociów** and **J Warycha**
Electrotechnical Institute, M. Sklodowskiej Curie 55/61, Wroclaw, Poland

E-mail: mielcar@iel.wroc.pl

Abstract. Tin oxide varistors may be an alternative for zinc based ones. The comparative analysis of both materials showed that, after adding suitable dopants, SnO_2 has similar semiconducting properties as ZnO. SnO_2 varistors were made in two versions: low- and high-added. The distinction lied in the added barium-bismuth compound in an amount of 0.3% and 0.8% respectively. Other metal oxides commonly used in varistors were added as well. Varistors were created using conventional ceramic technology applying sintering temperature 1250 °C.

1. Introduction

Some semiconductor oxides exhibit non-linear nature of the current-voltage relationship. This property has found practical application in varistors which in electrical circuits are elements serving to protect equipment against surges. ZnO varistors, due to their high coefficient of nonlinearity (approx. 50), high energy absorption, low leakage current and quick response to changes in voltage, are widely used as fuses in electrical appliances and as a surge arresters in the distribution and transmission power lines. SnO_2, due to its high electrical stability and less complex microstructure may be competitive to ZnO [1].

The properties that predispose ZnO for use in electrical engineering result from the fact that in its crystal lattice are interstitial atoms of zinc which cause that junctions of single crystals exhibit nonlinear electrical properties and their electrical properties can be easily modified using dopants. Not without significance is that zinc oxide elements can be easily prepared by conventional ceramic technology. The only disadvantage in the application of zinc oxide in varistors is a high conductivity of oxygen ions and low electrical stability at high currents.

When it comes to tin oxide it is characterized by a low degree of compaction during sintering, which predisposes it to be used in gas sensors [2] rather then in varistors. The poor compaction at low temperatures results from a low coefficient of diffusion of SnO_2 [3]. At higher temperatures (over 1273K) the process of sintering is worsened by the formation of the intermediate phases, a high partial pressure of SnO_2 vapors [4, 5] and the predominance of surface diffusion mechanisms over the evaporation – condensation ones. The evaporation - condensation (at high temperatures) processes cause only grain growth while tin oxide grains as such are not sintered. Sintering tin oxide can only be achieved through dopants, which are able to modify the crystal lattice of the SnO_2 grains [6].

The varistor effect in ZnO varistors, as well as in those based on SnO_2 ceramics is due to the transport of the majority carriers (electrons) by the potential barrier formed at the grain boundaries

[1] To whom any correspondence should be addressed.

IOP Conf. Series: Materials Science and Engineering **113** (2016) 012015 doi:10.1088/1757-899X/113/1/012015

during sintering in an oxidizing atmosphere [7]. Varistors based on SnO_2 as compared to the ZnO varistors are more resistant to degradation.

The phenomena of degradation of both SnO_2 and ZnO varistors were examined by [8], using two different approaches: The first method consisted of applying a DC voltage at elevated temperature and degrading with impulse 8/20 current (type lightning). The surface charge accumulated on the grain boundaries was analyzed before and after degradation using electrostatic force microscopy (EFM). As it appeared, prior to the process of degradation, in SnO_2 varistors the 85% of barriers were active, while the ZnO varistors contained only 30% of barriers that effectively participated in conduction by creating the semiconductor voltage barriers causing a drop of approx. 3V at 1mA. As a result of the degradation with pulses, the changes occurred in both cases. In the case of ZnO varistors, the current-voltage characteristics were essentially ohmic, due to the destruction of potential barriers (about 99% of interphase junctions were conductive). However, the SnO_2 system after the degradation with pulse currents continued to show no-ohmic behavior, although with a significant reduction in the amount of effective barriers (from 85% to 5%). In the case of aging with DC current, the SnO_2 system showed minimal differences in an amount the effective barriers before and after aging, whereas in the ZnO one, their number reduced from 30% to 5%. This result demonstrates that to the destruction of potential barriers and annihilation of defects much more energy is needed in the SnO_2 system than in ZnO one. Results obtained by [8] indicates that the SnO_2 varistors are suitable for high voltage applications and to that extent are potentially promising substitute for ZnO varistors.

2. Preparation of samples.

As materials for samples preparation the chemically pure (CP) chemicals were used. The samples compositions are given in Table 1.

Table 1. Chemical composition of varistors.

Symbol of sample	$BaBiO_{2.77}$	Sb_2O_3	Cr_2O_3	NiO	Co_2O_3	MnO	SnO_2	ZnO
WSn00	0,2	0,2	0	1	0,5	0,05	0	98,05
WSn05	0,2	0.05	0.6	1	0.5	0.15	0.5	97.84
WSn02	0,2	0.05	0	1	0,5	0,3	0,2	97,75
WSn98	0,3	0,05	0,05	1	0,5	0,15	97,83	0,05
WSn90	0,8	0,1	0,1	0,2	1	0,2	90,00	7,6

The mixtures of powders were then fabricated in varistors by forming in discs and sintering at 1250°C. The WSn00, WSn05 and WSn02 are the ZnO varistors with a small addition of SnO_2 while the WSn98 and WSn90 are the SnO_2 varistors, respectively low- and high-added with Ba-Bi compound. The samples were then subjected to microscopic and electrical examinations.

3. Experimental results

3.1. Microscopic characterization of samples

As can be seen in Figs 1 and 2, in the WSn00 sample there were no separate Bi or Ba phases while in the WSn05 sample there were some numbers of micron grains, which are $BaCr_2O_4$ grains in all probability.

IOP Conf. Series: Materials Science and Engineering **113** (2016) 012015 doi:10.1088/1757-899X/113/1/012015

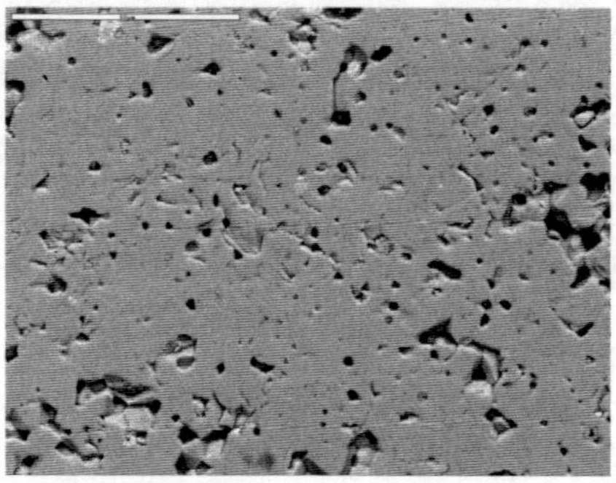

Figure 1. SEM view of microstructure of WSn00 sample

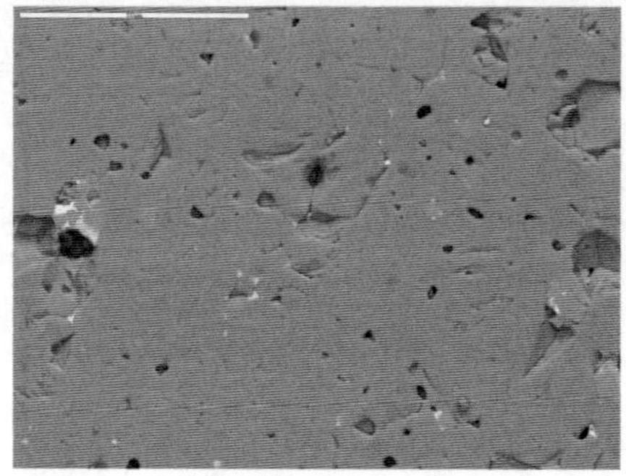

Figure 2. SEM view of microstructure of WSn05 sample

SEM observations of microstructures of WSn98 and WSn90 (Figs 3 and 4) varistor samples show homogeneous microstructures with evenly distributed bright spots, which are liquid phase of bismuth oxide. Due to it, in varistor body there are not areas excluded from work. The darker spots seen in the pictures are pores that were formed during polishing of the samples, whereby the tin oxide grains were washed out.

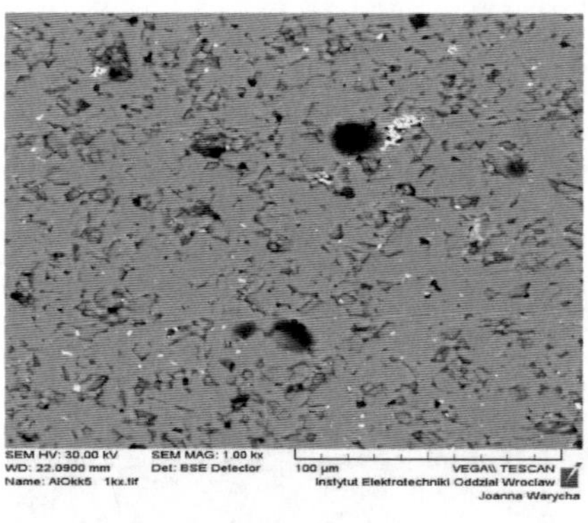

Figure 3. SEM image of microstructure of WSn98 sample, low-doped with Ba-Bi compound.

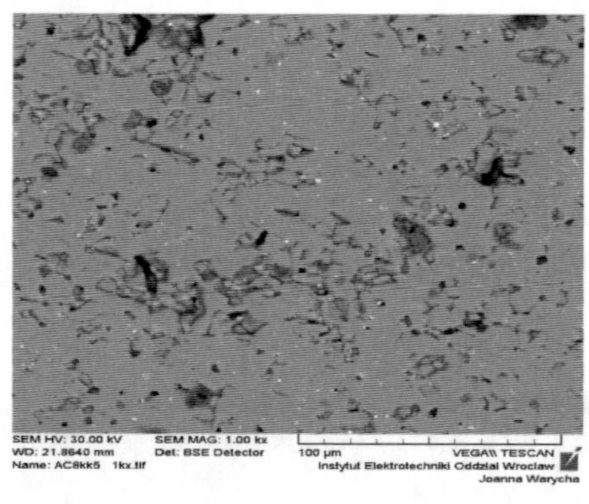

Figure 4. SEM image of microstructure of WSn90 sample, high-doped with Ba-Bi compound.

3.2. Elecrtical characterization of samples

The current-voltage characteristics were performed applying current in the range of 20 μA - 5 mA DC and a pulse current of 100 mS length, repeated every 1s. The I-V curves of both, ZnO varistors with small addition of SnO_2 and SnO_2 varistors are shown in Figs 5 and 6.

Varistor voltages of ZnO varistors with small addition of SnO_2 were 270 V/mm and 300 V/mm and nonlinearity coefficients α were about 50. Varistor voltages of WSn90 and WSn98 SnO_2 varistors were 270 V/mm and 320 V/mm respectively, while corresponding nonlinearity coefficients α were 6.4 and 4.4 respectively.

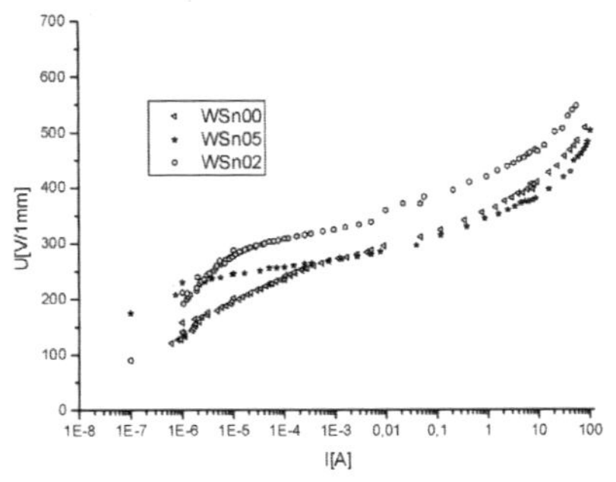

Figure 5. I-V curves of ZnO varistors with small addition of SnO_2

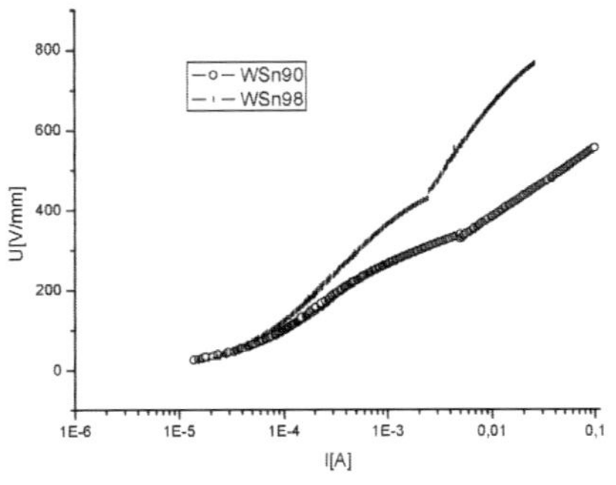

Figure 6. I-V curves of both low- and high-doped SnO_2 varistors.

4. Summary

Zinc oxide based varistors with small additions of tin oxide exhibited high (about 50) nonlinearity coefficients α and varistor voltages 270 and 300 V/mm. Similar varistor voltages have been exhibited by varistors made of tin oxide, but despite similar content of dopants, their α coefficients were significantly lower, 4.4 and 6.4. However, the microstructure of both types of varistors do not differ in a fundamental way. Added dopants were homogeneously distributed in bodies of both ZnO and SnO_2 varistors. The differences in I-V characteristics can be attributed to the different features of grain boundaries.

5. References

[1] Bueno PR, Varela JA and Longo E, 2007, SnO_2, ZnO and related polycrystaline compound semiconductors: An overview on the voltage dependent resistance (non-ohmic) feature, *J.Eur.Ceram.Soc*

[2] Pianaro S A, Bueno P R, Olivi P, Longo E and Varela J A, 1998, Electrical properties of the SnO_2 based varistors *Journal of Materials Science, Materials in Electronics* **159-65** 9

[3] Varela JA, Whittemore OJ and Longo E 1990 Pore size evolution during sintering of ceramic oxides *Ceramics International* **177-7916** (3)

[4] Mihaiu M S, Scarlat O, Zuca S and Zaharescu M, 1999 Advanced SnO_2-Based Ceramics: Synthesis, Structure, Properties *DOI: 10.5772/18211*

[5] Aguilar-Martınez JA, PechCanul MI, Hernandez MB, Glot AB, Rodrıguez E and Garcıa Ortiz L, 2013, Effect of sintering temperature on the electric properties and microstructure of SnO2–Co_3O_4–Sb_2O_5–Cr_2O_3 varistor ceramic *Ceramics International* **4407–12** 39

[6] Cássia-Santos MR, Sousa VC, Oliveira MM, Sensato FR, Bacelar WK , Gomes JW, Longo E, Leite ER and Varela JA 2005 Recent research developments in SnO_2 based varistors, *Materials Chemistry and Physics* **1–9** 90

[7] [Glot A B, Lu Z Y, Zhou Z Y and Ivon A I 2011 Electrical properties of SnO_2 ceramic varistors withstanding high current pulses *Superficies y Vacío* **61-7** 24(2)

[8] Ramírez M A, Tararam R, Simões AZ, Ries A, Longo E and Varela J A, 2013, Degradation Analysis of the SnO_2 and ZnO-Based Varistors Using Electrostatic Force Microscopy, *Journal of the American Ceramic Society* **1801-09** 96 (6)

IOP Conf. Series: Materials Science and Engineering **113** (2016) 012016 doi:10.1088/1757-899X/113/1/012016

The investigation of the light radiation caused polyethylene based materials deterioration by means of atomic force microscopy

A Sikora[1], A Grabarek[2], L Moroń[1], M Wałecki[1], P Kryla[1]
[1]Electrotechnical Institute, Division of Electrotechnology and Materials Science, M. Skłodowskiej-Curie 55/61, 50-369 Wrocław, Poland
[2]Faculty of Physics, Astronomy and Applied Computer Science and Faculty of Chemistry, Jagiellonian University, prof. Stanisława Łojasiewicza 11, 30-348 Kraków, Poland

E-mail: sikora@iel.wroc.pl

Abstract. The impact of the environmental conditions on the materials used in various devices and constructions, in particular in electrotechnical applications, has an critical impact in terms of their reliability and utilization range in specific climatic conditions. Due to increasing utilitarian requirements, technological processes complexity and introducing new materials (for instance nanomaterials), advanced diagnostic techniques are desired. One of such techniques is atomic force microscopy (AFM), which allows to study the changes of the roughness and mechanical properties of the surface at the submicrometer scale, enabling the investigation of the degradation processes. In this work the deterioration of selected group of polyethylene based materials have been measured by means of AFM, as the samples were exposed to the simulated solar light and UV-C radiation. Such an analysis of the environmental conditions impact on the deterioration process using AFM methods for various versions of specific material was not presented before.

1. Introduction

The issue of the reliability and utilization economy of various products, is one of the major concerns of each consumer. A number of works aimed at improvement and verification of those factors have been performed [1-3]. The major impact on the product's performance, including the lifetime is the durability of used materials. The continuous progress in material development requires verification of their various properties, including the resistance to the environmental conditions, as one of the essential factors in terms of outdoor applications. Due to the need of the insight into the decomposition of tested materials and nanomaterials in particular, advanced diagnostic methods can provide useful information. Such an approach, based on utilization of atomic force microscopy, was already presented in various works, aimed at observation of the nanoscale impact of light radiation, increased temperature or high voltage on the surface of tested samples [4-10]. Also, the molecular-scale wear of the material due to repetitive load could be observed [11]. In addition, specific methodology solutions

[1] To whom any correspondence should be addressed.

were developed in order to improve the analysis quality [12-15]. However, systematic works, providing step-by-step observation of the material's deterioration are so far rare.

In this work, the investigation of three types of polyethylene based materials are presented. The impact of the simulated solar and UV-C radiation was verified in terms of the roughness and stiffness changes. Moreover, the wettability of the surface, as the roughness-related phenomena was measured. Obtained results allowed to notice various resistance of tested materials to the applied conditions, as the added compounds improve their properties. In addition, one can conclude various dynamics of the surface degradation. Moreover, the force spectroscopy measurements provided unique data allowing to identify both: detachment of the microparticles and the appearance of the electrostatic charge embedded in the material, related to the polymer chains breaking. Therefore, the superior role of atomic force microscopy in the investigation of the materials deterioration process was shown.

2. Experimental results

2.1. The preparation of the samples and the exposition the light radiation

Three materials have been prepared in order to perform the experiment: all materials are based on the linear low density polyethylene (LLDPE – DFDA 7540 from Dow Chemical Company). The material marked "PE" is the pristine polymer. The composite "PE+UV" contains the UV-stabilizer (UV 6014 LD from Polyplast Muller) in amount of 5 wt. %. The "PE+UV+MMT" was filled with the UV-stabilizer and the montmoryllonite (Cloisite 15A from Southern Clay). The composite PE-UV was prepared in a single-screw extruder. The technology of PE-UV-MMT was a two-stage process. First the montmoryllonite was introduced into polyethylene matrix in a twin-screw extruder, than this material was homogenized with the UV stabilizer in the single-screw extruder. The samples of the two composites, as well as the pristine polymer were formed by press moulding by 40 tons at 150 ^{0}C for 6 min. The samples were placed in two independent chambers providing simulated solar radiation emitted by the xenon lamp, and UV-C radiation emitted by evanescent lamp. Following basic parameters of the radiation conditions were applied: Xenon lamp (Xe): λ_{range}=300-800 nm and 38,5 W/m^2, UV lamp (UV): λ=235 nm and 9 W/m^2. In order to observe the deterioration progress, the following time periods were applied: 100, 200 and 500 hours.

2.2. AFM topography analysis

The morphology analysis was performed using DI 3000 AFM, working at ambient conditions, in tapping mode. To perform the measurement, following probes were used: *NanosensorsPointprobes*, f_{res}=306-353 kHz, k=43-68 N/m. Scanning area was 3μm x 3μm (fig. 1a, b), as it provided desired resolution allowing to observe nanometer-scale features, and avoid the influence of micrometer/milimeter-scale objects related to the form imperfections (fig. 1c).

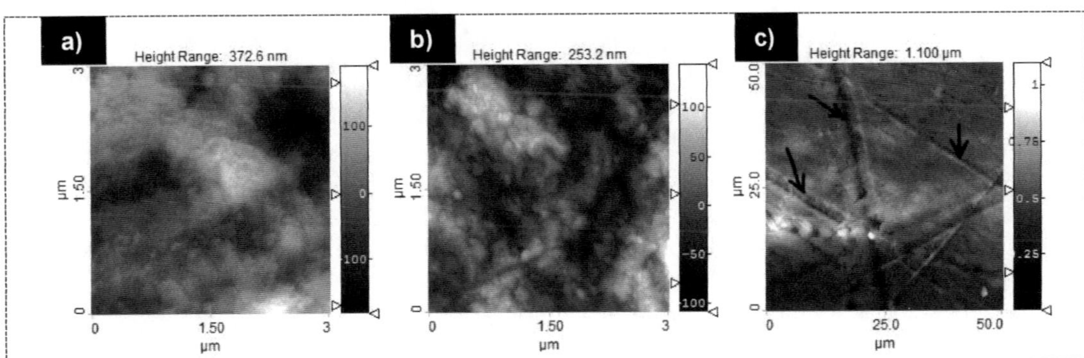

Figure 1. The examples of the surface scans 3μm x 3μm (a, b) and 50μm x 50μm (c).

In order to obtain statistically relevant data, at least 20 scans for each case were acquired. For each image, the roughness parameters were calculated, including Sq (Root Mean Square) and Sbi (Surface Bearing Index). The median, second and third quartiles were calculated. The data was processed using SPIP software from Image Metrology company [16]. No data revealing the influence of the fabrication process like presented in figure 1c was taken into account. Obtained results are presented in figure 2.

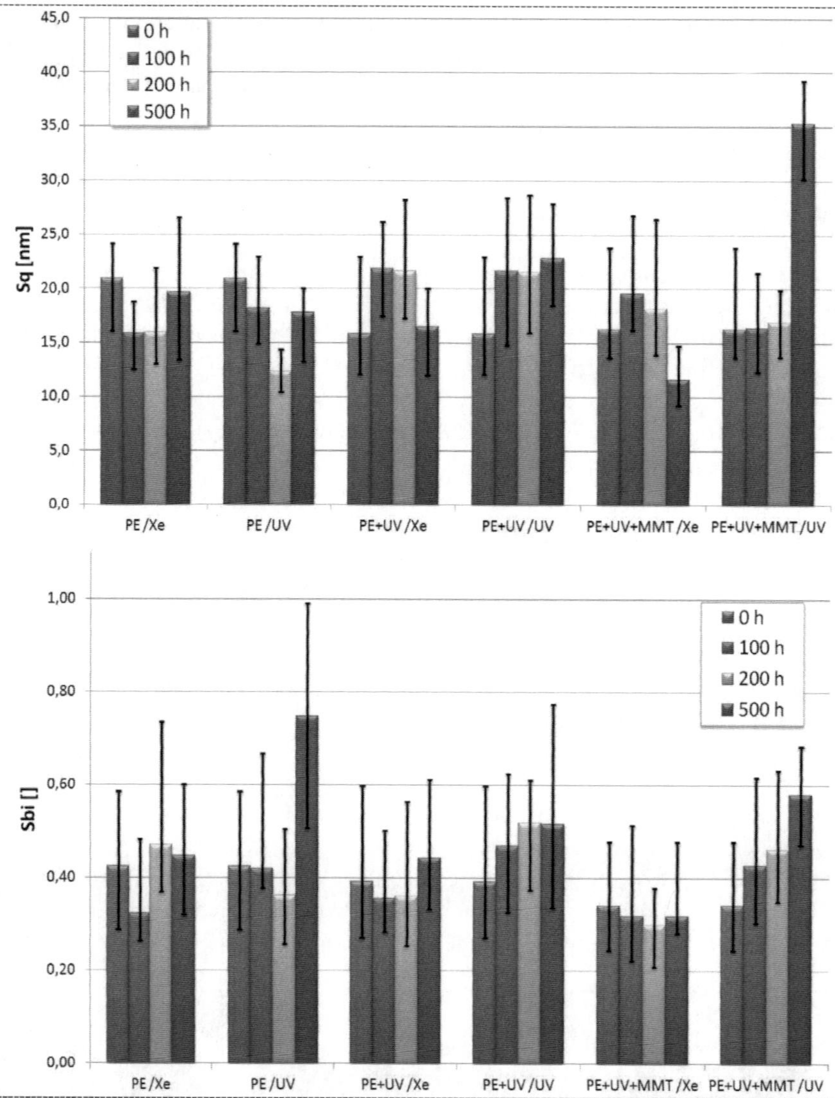

Figure 2. Roughness parameters (Sq and Sbi) changes of investigated samples.

Despite relatively large distribution of analyzed factors caused by the surface non-homogeneities, one can observe specific changes of the roughness. The Sq parameter reveals non-monotonic behavior, where at the first stages the roughness decrease is observed, and when the deterioration process continues, the increase occurs. Such a phenomena may seem surprising, however it was observed before, when the polycarbonate sample was tested in terms of the UV light resistance [13]. One can argue, that the first parts of the surface that deteriorates, are elevated, exposed and fragile areas. Once they are gone and Sq reaches minimum, the holes start to appear due to further degradation, and the roughness increase. One has to be aware, that in case of presence of any fillers, such a process may show different progress tendency. Obtained results allowed to observe exceptions from general behavior, as samples PE+UV/Xe and PE+UV+MMT/Xe aged using Xenon lamp reveal reversed tendency. Such a phenomenon could be caused by a presence of UV stabilizer, which increased the resistance of the material against the radiation, however the impact of the thermal interaction could lead to some level of surface's degradation. However, in following steps, the surface become more flat with the majority of holes (see Sbi discussion below). Rapid increase of Sq in case of PE+UV+MMT sample irradiated with UV, may be related to the MMT particles uncovering, which as relatively big objects may have a major contribution in the roughness changes.

The Sbi parameter, which is related to the bearing properties, in general increases, therefore one can conclude, that the upper part of the surface flattens continuously, however the amount and size of holes increase (also indicated by Ssk parameter not shown here). In addition, one can also notice general behavior, where the UV light causes larger relative changes of the surface roughness due to its higher energy, than Xenon lamp light. It should be also stressed, that the presence of the additives did not influence significantly the surface roughness. Po particular improvement of the roughness changes was observed after UV stabilizer added.

2.3. AFM surface stiffness determination

The force spectroscopy measurements allowed to determine the changes of the surface stiffness. *CSG30* probes from *NT-MDT,* f_{res}=48 kHz k_{nom}=0.6 N/m. In order to calculate the real spring constant ,thermal noise method was used. For each case approximately 300 force curves have been acquired (10 in single spot) in order to provide statistically reliable outcome. Figure 3 shows the examples of acquired curves, revealing unique phenomena such as the microscale cracking (a) of the surface due of the applied force, or appearance of the electrostatic force due to the decomposition of the polymer chains (b). It should be stressed, that no other measurement technique allows to obtain such a results. Further investigations performed by means of electrostatic force microscopy (EFM) may provide valuable information concerning the way this phenomena develops. It should be underlined, that the electrostatic forces mapping provided by EFM allows to detect locally accumulated charge in insulating materials [17, 18].

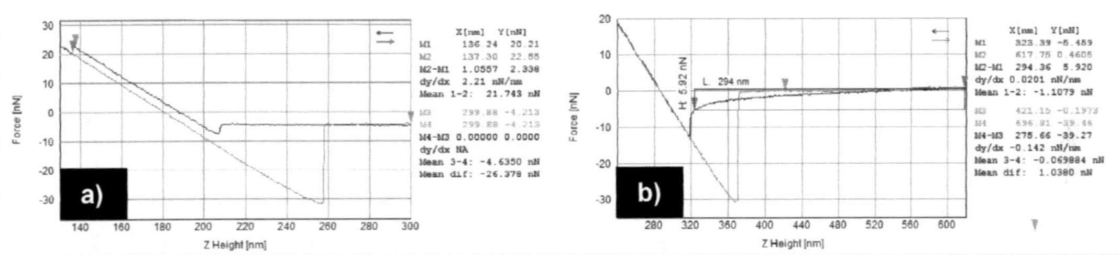

Figure 3. The examples of the force spectroscopy curves revealing the surface cracking under the load (a) and the presence of electrostatic force (b).

Figure 4 shows the stiffness changes of investigated materials. This parameter was determined basing on the slope of the tip-sample interaction in repulsive forces range. Acquired data was processed in order to obtain statistical information about the average value and the median.

Obtained results allow to observe complex, non-monotonic changes of the mechanical properties of the surface. At the first stage the stiffness increases. This particular phenomena can be related to the deterioration of relatively soft, elevated features, weakly attached to the surface. Following, as the abovementioned features are gone and the degradation process continues, the material's surface stiffness decreases due to the activation of the radicals and the molecular chains decomposition.

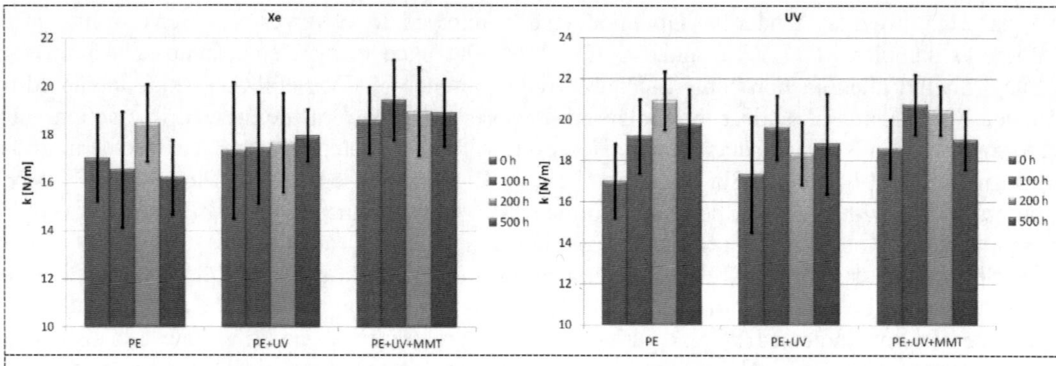

Figure 4. The changes of the stiffness of investigated materials.

The differences in the process dynamics can be seen between certain samples and exposition conditions. All samples exposed to UV light show reversed "V" shape, revealing both phases of the process. One should be aware, that in case of Xenon lamp exposition, deep deterioration of the material, caused by the raised temperature, can't be detected straightforward using force spectroscopy tool, however, the presence of the material cracks or discontinuities, may be revealed as the tip-sample interaction events in the repulsive force range. As expected, the dynamics of the stiffness changes is higher for the UV light exposition. It can be also seen, that the addition of MMT provided higher stiffness, allowing to obtain better durability of the material.

2.4. Wettability angle determination

In addition to the roughness data, the wettability angle was determined, as it also plays essential role in the degradation of the material in outdoor applications. The measurements were performed using the home-made setup containing CCD-camera based microscope and digitally controlled deionized water dispenser. For each sample at least 10 angles have been acquired in order to provide statistical analysis. The graphs showing the wettability changes are presented in figure 5.

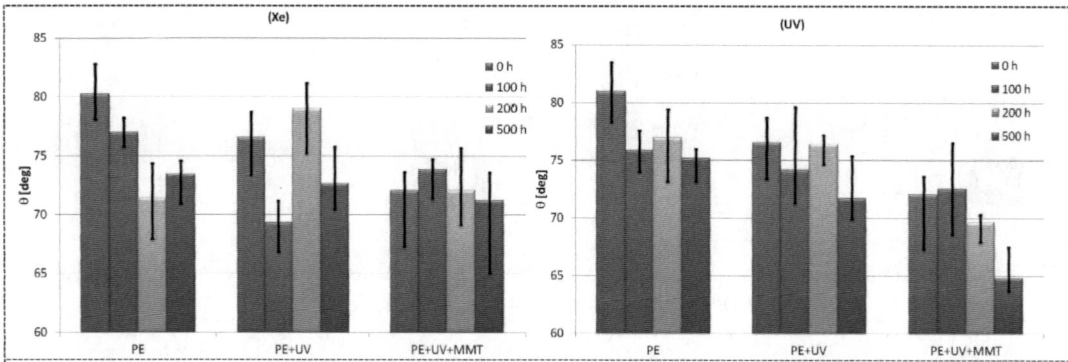

Figure 5. The wettability angle changes due to the deterioration of the surface's of investigated materials.

The wettability angle, as expected, decreases along with the radiation period, proving the roughness increase of the investigated surface. Therefore, in general this result confirms the roughness increase. Such a technique, however, due to the presence of milimeter scale features related to the fabrication process (fig. 1c), can not provide such an insight into the morphology changes as the AFM methods deliver. Again, the UV radiation caused more dynamic changes of measured factor than Xenon lamp light. It should be underlined, that UV stabilizer may present a limited performance in terms of presence of UV-C radiation. There must be stressed, however, that the study of the resistance of materials to such a wavelength are useful, as certain object can be exposed to such conditions for instance in medical facilities.

3. Summary and outlook

Performed experiments allowed to trace the changes of the morphological and mechanical properties of the PE and PE based samples at the submicrometer scale. Acquired data delivered unique information about the deterioration of the investigated materials due to exposition to both: Xenon and UV-C light sources. The multi- stage process in both: roughness and stiffness changes was observed. Observed degradation phenomena contained UV radiation curing and polymer chains breaking, preliminary surface smoothing and the then the holes appearance. Also, the impact of presence of UV stabilizer and MMT nanoparticles was verified. It should be stressed, that no such complex study was performed so far. Additionally performed wettability tests provided confirmation of the surface's roughness development.

Further research will be performed in order to obtain more detailed information about various aspects of the materials deterioration. Long term tests may provide the better insight into the process in terms of understanding the role of various nanofillers. In particular, the detailed study of the appearance of the electrostatic forces will be carried out, using the electrostatic force microscopy technique. It should be stressed, that the presence of the electrostatic field may have a considerable consequence in terms of the ageing process as well as the aggravation of the material's performance as the insulator. Presented investigation are particularly valuable in terms of observation of the behaviour of the nanomaterials, where the submicrometer scale parameters can deliver essential information.

4. References

[1] Falat T, Matkowski P, Zaluk Z, Felba J and Moscicki A 2014 *Proceedings of the 2014 37th International Spring Seminar on Electronics Technology*, ISSE 2014, **6887626**, 365-368.
[2] Szeloch RF, Janus P, Serafińczuk J, Szecówka PM and Jóźwiak G 2012 *Microelectronics Reliability*, **52** (4), 711
[3] Sikora A, Kryla P, Wałecki M and Bednarz Ł 2013 *Przegląd Elektrotechniczny* **89,** 336
[4] Arefian M, Zia M, Tahmourespour A and Bayat M 2013 *Int. J. Environ. Sci. Technol.* **10,** 1319
[5] Berthumeyrie S, Colin A, Esparcieux C, Baba M, Catalina F, Bussiere PO and Therias S 2013 *Polym. Degrad. Stab.* **98,** 2081
[6] Collin S, Bussière PO, Thérias S, Lambert JM, Perdereau J and Gardette JL 2012 *Polym. Degrad. Stab.* **97,** 2284
[7] Gu X, Michaels CA, Nguyen D, Jean YC, Martin JW, Nguyen T 2006 *Appl. Surf. Sci.* **252,** 5168
[8] Jonda C H, Mayer A B R and Stolz U 2000 *J. Mater. Sci.* **35,** 5645
[9] Larché J F, Bussière P O, Thérias S, and Gardette J L 2012 *Polym. Degrad. Stab.* **97,** 25
[10] Nguyen T, Martin J, and Byrd E 2003 *J. Coatings Technol.* **75,** 37
[11] Sikora A, Iwan A 2012 *High Performance Polymers* **24**(3) 218
[12] Sikora A 2013 *Optica Applicata* 43(1), 163
[13] Sikora A 2014 *Meas. Sci. Technol.* **25,** 055401
[14] Proff C, Abolhassan S, Dadras MM and Lemaignan C 2010 *Nucl.Mater.* **404** 97
[15] Rust MA and Todd RI 2007 *Mater. Sci. Forum* **551–552** 615
[16] http://www.imagemet.com/WebHelp6/Default.htm#RoughnessParameters/Roughness_Parameters.htm accessed 25.10.2015
[17] Sikora A 2009 *Acta Phys. Pol.* **A 116**, 102
[18] Bonilla R, Ávila A, Montenegro C, and Hinestroza J 2012 *J. Microsc.* **248**, 266

Acknowledgments
Presented experiments were performed within frames of IEL statutory work.

Corrosion resistance of the AISI 304, 316 and 321 stainless steel surfaces modified by laser

B. Szubzda[1], A. Antończak[2], P. Kozioł[2], Ł. Łazarek[2], B. Stępak[2], K. Łęcka[2], A. Szmaja[1], M. Ozimek[1]

[1]Electrotechnical Institute Division of Electrotechnology and Materials Science, M. Skłodowskiej-Curie 55/61, 50-369 Wrocław, Poland,
[2]Institute of Telecommunications, Teleinformatics and Acoustics, Wroclaw University of Technology, Wybrzeże Wyspiańskiego 27, 50-370 Wrocław, Poland.

corresponding author: b.szubzda@iel.wroc.pl

Abstract: The article presents the analysis results of the influence of laser fluence on physical and chemical structure and corrosion resistance of stainless steel surfaces modified by irradiating with nanosecond-pulsed laser. The study was carried out for AISI 304, AISI 316 and AISI 321 substrates using Yb:glass fiber laser. All measurements were made for samples irradiated in a broad range of accumulated fluence (10÷400 J/cm^2). The electrochemical composition (by EDX) and surface morphology (by SEM) of the prepared surfaces were carried out. Finally, corrosion resistance was analyzed by a potentiodynamic electrochemical test. The obtained results showed very high corrosion resistance for samples made by fluency of values lower than 100 J/cm^2. In this case, higher values of corrosion potentials and breakdown potentials were observed. A correlation between corrosion phenomena, the range of laser power (fluence) and the results of chemical and structural tests were also found.
Keywords: Stainless steel, laser marking/engraving, corrosion resistance, electrochemical test, salt spray test.

1. Introduction

The main advantages of using laser in materials engineering are non-contact working, high repeatability, high scanning speed, worked area comparable to the laser spot, high flexibility and automation. In the last years, one of the developed laser application has been marking/engraving. Laser marking/engraving allows for labelling various kinds of patterns or texts on metallic and plastic products or component surfaces [1,2] and is widely applied in, for example, making product serialization, copyright patterns, automotive industry, surface textures or gifts and metal jewellery design [3]. In the case of metallic materials, the laser application in the surface treatment is focused on, among other things, aluminium and titanium [4]. Many applications refer to different kinds of stainless steels [5].

When a metallic surface is irradiated with high power laser radiation, one part of energy is reflected and the other is absorbed. The absorbed part rapidly increases the subsurface temperature which can cause the surface microstructure and composition modification, oxide film forming, crystal defect initiating, changing in crystal structure or phase transition [6,7]. In the case of stainless steel, it can be a serious problem due to the possibility of corrosion resistance change and thereby future exploitation properties.

The aim of this work was to evaluate the effect of laser treatment on the corrosion resistance of stainless steels, such types as AISI 304, AISI 316 and AISI 321. The study was performed for the commonly used MOPA Yb:glass fiber laser (1062 nm) configuration. The electrochemical composition (by EDX) and surface morphology (by SEM) of the obtained surfaces were carried out; corrosion resistance was analyzed by a potentiodynamic test – according to Norm [8].

2. Materials

Three types of stainless steel (AISI 304, AISI 316, AISI 321) with different chemical compositions were selected (Table 1). In spite of inconsiderable differences in the content of alloy additions, they are characterized by different corrosion resistance and are used in various applications. Stainless steel AISI 304 is used most frequently, which results from its anti-corrosion properties as compared to the price. AISI 304 can be used in the atmospheric environment, natural water, alkaline solutions and certain acids. The most popular acid-resistant material is steel AISI 316, which exhibits increased resistance to corrosion in very aggressive chemical environments. The third tested type is AISI 321, which shows high resistance to inter-crystalline corrosion and resistance to the work in higher temperatures, i.e. 400 – 800°C [9].

Table 1. The percentage chemical composition of stainless steels

	C	Mn	Si	P	S	Cr	Ni	Mo	Ti
AISI 304	max 0,08	max 2,00	max 1,00	max 0,045	max 0,030	18,00 20,00	8,00 10,50	-	-
AISI 316	max 0,08	max 2,00	max 1,00	max 0,045	max 0,030	16,00 18,00	10,00 14,00	2,00 3,00	-
AISI 321	max 0,08	max 2,00	max 1,00	max 0,045	max 0,030	17,00 19,00	9,00 12,00	-	max 0,70

From each of the three types of steel, 60x100 mm rectangular samples were cut out of 1 mm sheets. After rinsing the surface with isopropyl alcohol, the test layers were made by means of laser radiation with fluences in the range in which it was possible to obtain laser colourful coating. The value of fluence, which is used to determine the degree of intensity of laser radiation impact, takes into account individual parameters of the laser influence in accordance with the accumulated fluence formula [4].

$$F = \frac{2\sqrt{2}P}{\pi \varpi_0 V} \quad [\mathrm{J/cm^2}]$$

where P – laser power (W), ω_0 – beam waist of the laser (microns) and V – scanning rate (mm/sec). The laser power was constant and was 4.4 W and a change in fluence values was achieved by changing laser scanning rates of sample surfaces. For each of the three tested steels, four series of colourful layers with eight fluence values which were changed by 50 J/cm^2 in the range from 50 J/cm^2 to 400 J/cm^2 were carried out.

3. Experimental details

Each of the four test series was subjected to other tests which checked the properties of the obtained layers: a chemical composition and morphology, electrochemical corrosion tests and corrosion resistance tests in neutral salt mist.

3.1. Chemical composition testing

The measurement was made with the use of scanning electron microscope (SEM) Zeiss EVO 25 with an extension enabling an additional analysis of the elemental composition using X-ray spectroscopy with energy dispersion (EDX). Measurements on 300x300 µm surfaces were carried out, which made it possible to become independent of local defects of the surface morphology, recognizing the measuring field as representative for the whole test field. The percentage composition of the researched elements was defined. The fact that the depth of beam penetration for the X-ray analysis reaches up to 1 μm, which may exceed the thickness of the produced surface layers, was taken into account when interpreting the results [10,11].

The researched continually have an austenitic structure with an interstitial solid solution of one or more elements from Fe-γ. The main components of these alloys are chromium and nickel. It is the content of chromium that predominantly determines the anti-corrosion steel properties because the oxides of this metal form a passive protection layer, whereas the remaining components stabilize this layer in different situations resulting from the work environment. Figure 1 presents a comparison of the content of individual elements in each of the three types of the researched steels.

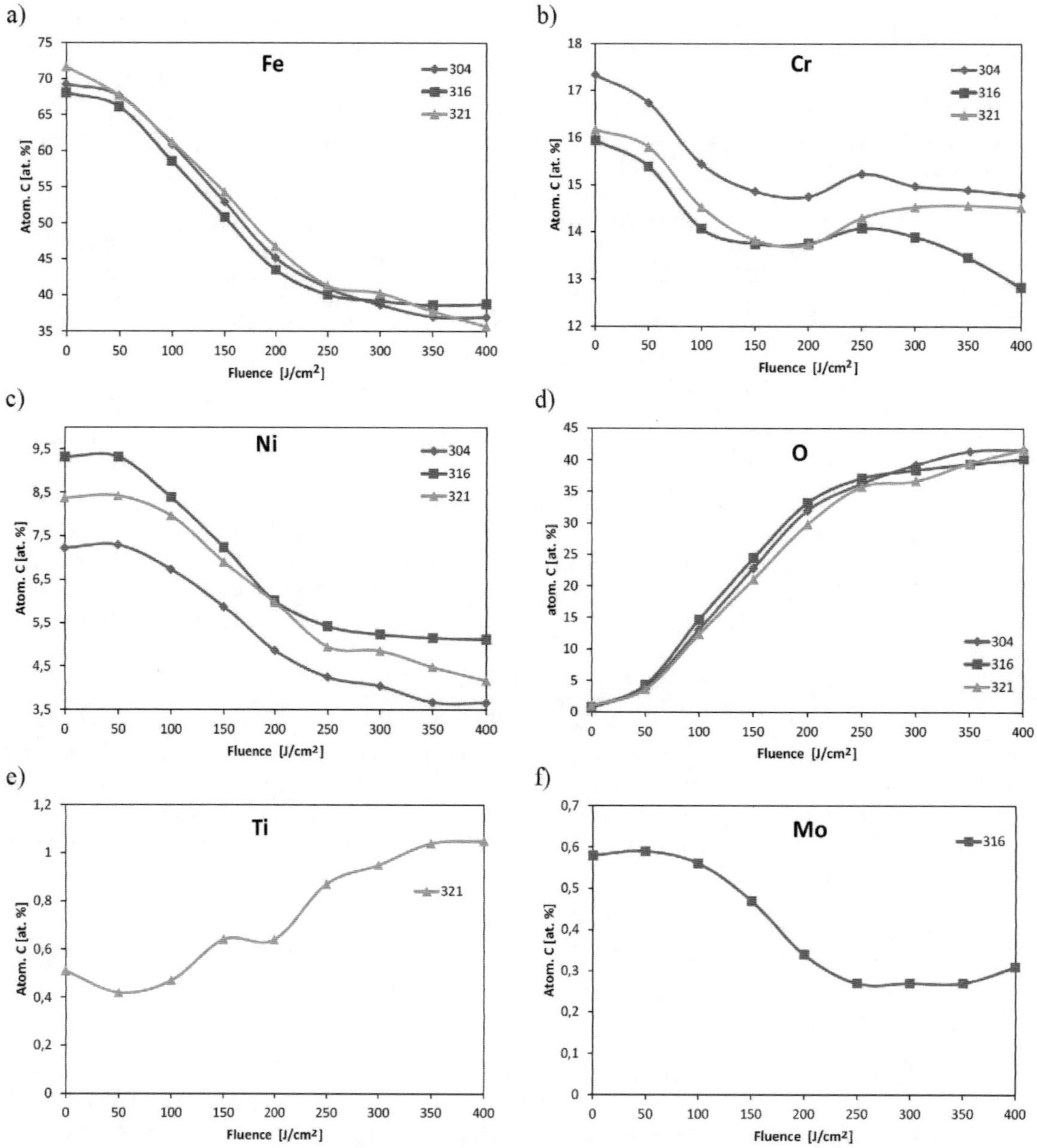

Figure 1. The measurement results of the each element content in the surface layer on three tested types of steel.

The content of oxygen in the subsurface layers, which grows along with the increase of fluence, shows a formation of oxides of alloy metals included in the steel, hence a similarly decreasing percentage content of iron and nickel. During laser heating of the surface, a disarrangement of the outer-self-passivated coating consisting mainly of Cr_2O_3 occurs because it is not stable in high temperatures, hence different changes in the content of chromium. Chromium atoms diffuse into the subsurface clusters, whose depth and multiplicity in this case depend on the temperature generated by a laser beam [12].

3.2. Surface morphology

During the process of laser heating of metal surfaces, oxidation reactions of part of the metallic alloy components occur in the surface layer. The oxide layer may vary as to the chemical composition, thickness, continuity and adhesion to the substrate depending on the laser power (fluence) and the generated temperature. All changes will significantly affect corrosion resistance of the layer.

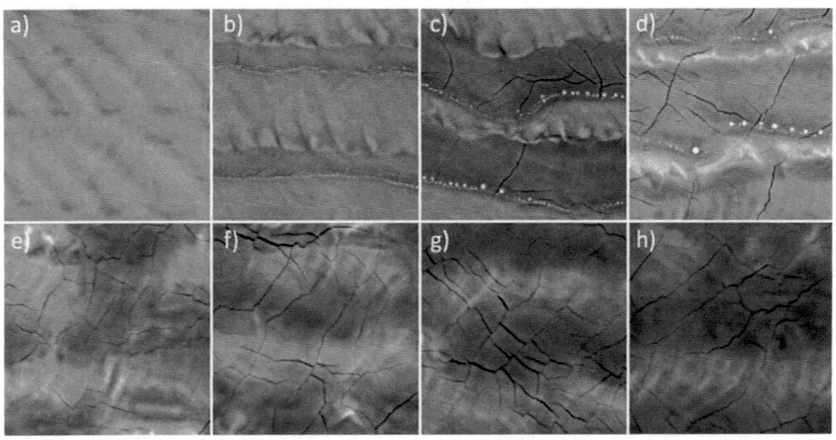

Figure 2. Microscopic photographs of scale surfaces produced on AISI 304 steel by means of radiation with the following fluence: a) 50 J/cm^2, b) 100 J/cm^2, c) 150 J/cm^2, d) 200 J/cm^2, e) 250 J/cm^2, f) 300 J/cm^2, g) 350 J/cm^2, h) 400 J/cm^2

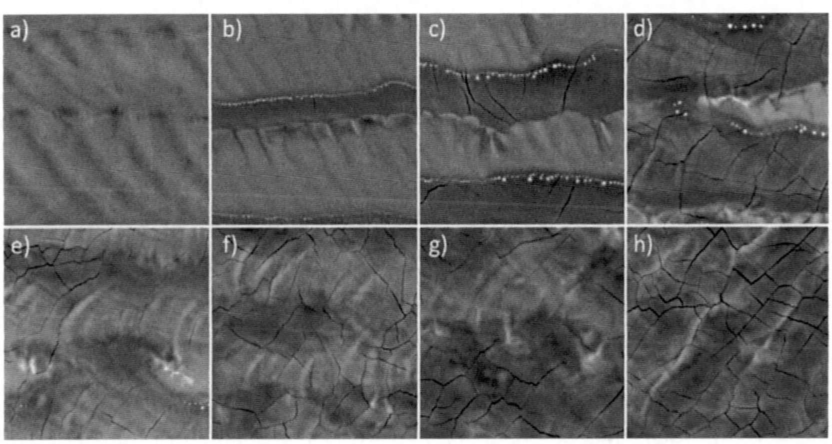

Figure 3. Microscopic photographs of scale surfaces produced on AISI 316 steel by means of radiation with the following fluence: a) 50 J/cm^2, b) 100 J/cm^2, c) 150 J/cm^2, d) 200 J/cm^2, e) 250 J/cm^2, f) 300 J/cm^2, g) 350 J/cm^2, h) 400 J/cm^2

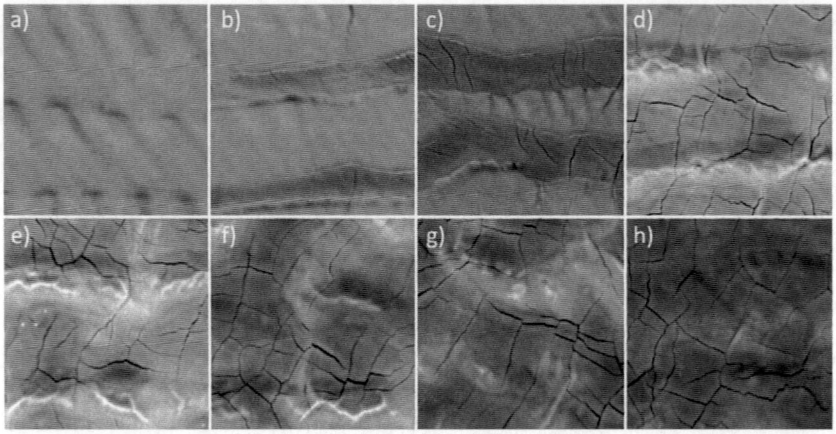

Figure 4. Microscopic photographs of scale surfaces produced on AISI 321 steel means of radiation with the following fluence: a) 50 J/cm^2, b) 100 J/cm^2, c) 150 J/cm^2, d) 200 J/cm^2, e) 250 J/cm^2, f) 300 J/cm^2, g) 350 J/cm^2, h) 400 J/cm^2

The drawings present images of external surface fragments of the examined layers measuring 20 × 20 microns and at a magnification of 4000x which were recorded by means of a scanning electron microscope (SEM) Zeiss EVO MA 25.

The surface of the layer formed at a fluence of 50 J/cm^2 is continuous and tight on each tested steel. It can be clearly seen that starting from a value of 100 J/cm^2 more and more micro-cracks occur on the passive oxide layer along with the increase of fluence. This phenomenon is similar for all analyzed steels.

3.3. Corrosion resistance

The tests were performed by a potentiodynamic method in accordance with Norm [8].

The electrolytic environment of measurements was chosen in such a way that it could reflect the most frequent risks in the conditions of a potential application of layers. The tests were carried out in a diluted solution of sulfuric acid having a pH of 3,5. The range of variability of measuring potentials for each of the solutions was from -1 V to +0,8 V against a saturated measuring electrode. The measurements were made in a flat measuring cell – type A-011951, where a counter electrode was a platinum wire. Changes in potentials and the measurement of intensities caused by currents were carried out by means of potentiostat – galvanostat ATLAS 0531 device. Changes in potentials occurred at a rate of 1 mV/s.

The measurement results were compiled by means of Bio-Logic Company computer program EC-LAB version 10.37 – Science Instruments. For each measurement, on the basis of anode and cathode coefficients of the slope of straight lines which obeys the Tafel's law, the following process parameters of the corrosion process were specified: corrosion potential Ek [mV], corrosion current density Jk μA/cm^2.

The results presented below were calculated by taking into account a potential of the calomel electrode. By increasing the potentials further over the values of the equilibrium potentials of Tafel cathode and anode processes, the potentials values of layer breakdown Ebreakdown [mV] were specified, in which a sudden increase in the intensity of the measured currents occurred.

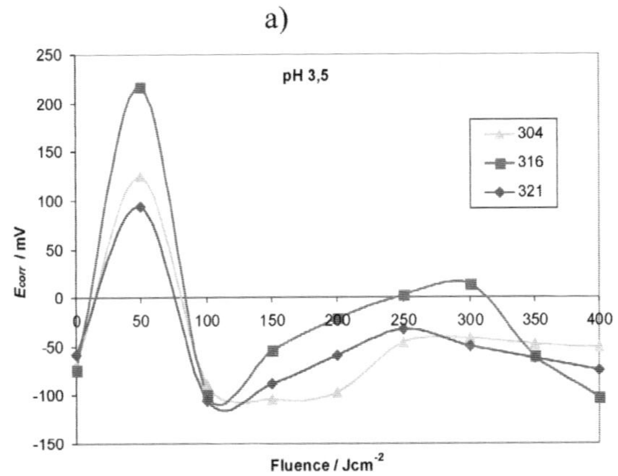

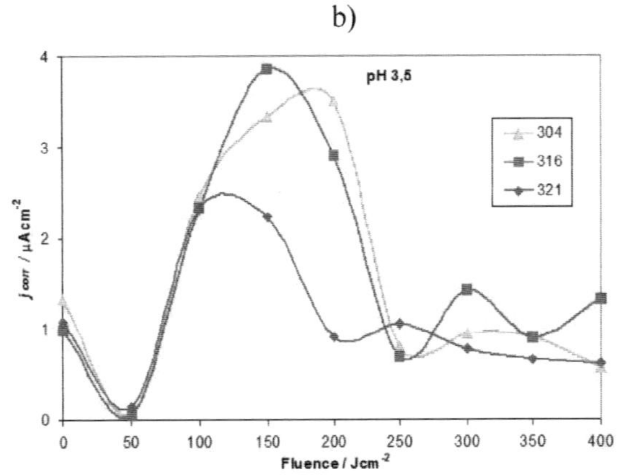

Figure 5. Results of potentiodynamic electrochemical tests, dependencies of corrosion potentials and corresponding current densities on fluence of laser radius impact. The point marked on charts as a fluence value of 0 is characterized by corrosion susceptibility of steels which did not undergo laser treatment.

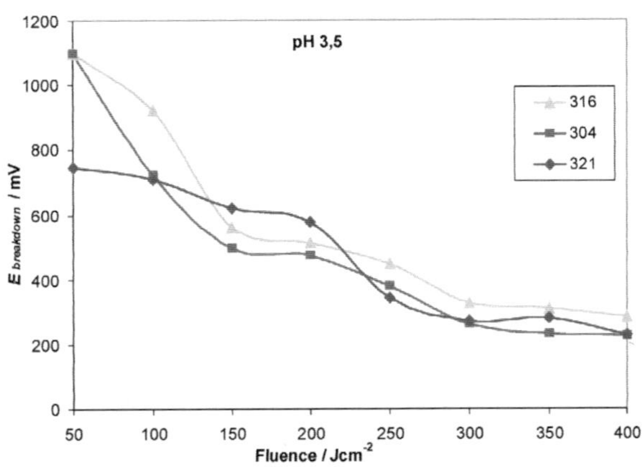

Figure 6. Results of potentiodynamic electrochemical tests, dependencies of layer breakdown potentials on laser radius impact fluence

The obtained results should be considered to be very interesting. The continuity of layers produced on each of the researched steels at a laser radiation fluence of 50 J/cm^2, which was found in Fig. 2–4, has a particularly significant confirmation in the result of the corrosion potential. The increased corrosion resistance of these layers in relation to the very material of the substrate (fluence 0) reaches very high positive values, for example, for untreated steel 304 E_{corr} it is circa -80 mV and with a layer produced at a fluence of 50 J/cm^2 it reaches the values which are characteristic of a very chemical-resistant precious metals of circa 220 mV (Fig. 5a). This proves not only the continuity of the layer but also its tightness, adhesion as well as its chemical and physical structure, which cannot be stated only by reference to the elemental composition – Fig. 1. The layers produced for the subsequent fluences of laser radiation are not continuous, which is confirmed by observations – Fig. 2-4. The values of their corrosion potentials fall rapidly to the value of bare steel and even below. Particularly noteworthy is the monotonicity of changes in the corrosion resistance of layers on each steel and for fluence values exceeding 50 J/cm^2, it does not occur similarly to the degree of surface cracking Figs. 2-4, as it might be expected, but in a very similar way as it is in the case of the subsurface content of chromium Fig.

1b. This means that the cracking degree of layers, width and the number of gaps is of secondary importance because the moment the electrolyte has contact with the substrate along with the increase in anode potentials, it generates corrosion processes which are controlled only by corrosion resistance of the substrate, and in this case it mainly depends on the content of chromium bonded in the oxide passivation [12]. However, this conclusion is not confirmed in the measurement results of breakdown potentials Fig. 6. We can observe a constant decrease in resistance onto layer breakdown along with fluence, which clearly corresponds to observations as regards continuity of layers Figs. 2-4. Therefore, there is no equal access of the electrolyte to the substrate for each of the layers with fluence higher than 50 J/cm^2 as it could be concluded from the above considerations.
The course of variability of corrosion current intensities in the fluence function is also interesting. The corrosion potential identifies areas of particular processes occurrence, whereas the corrosion current, especially its density, determines the intensity of individual processes. The corrosion rate is determined on the basis of this value. Observing the curves in Fig. 5b, we can see a confirmation of very high resistance of chemical layers formed at a fluence of 50 J/cm^2; current densities are minimal there and there is almost no flow of charges, above 50 J/cm^2 corrosion currents grow rapidly reaching very high values, which we note until a fluence of 250 J/cm^2 only in the case of steels 321 to 200 J/cm^2. Above these values the range of corrosion currents corresponds to values of steels which were not treated with laser. This representation slightly deviates from monotonicity of changes in corrosion potentials. Following the previous considerations, we can therefore discuss the following conclusion: there are three areas of dependencies of corrosion processes intensity on laser radiation fluence producing surface layers such as:
1. The area of the fluence value up to 100 J/cm^2, very high corrosion resistance resulting from peculiar properties of the passivation layer, maximum of resistance at 50 J/cm^2
2. The area of the fluence value above 200-250 J/cm^2 due to the discontinuity of outer layers, the resistance of the substrate material decides about corrosion resistance taking into account the fact that as a result of thermal effect of the laser beam in the subsurface layer, a change in the chrome content took place – which determines resistance Fig. 1b.
3. The area from 100 J/cm^2 to 200-250 J/cm^2 – a mixed influence of factors determining corrosion resistance in both neighboring areas. Resistant surface layers are discontinuous and do not protect the material, the availability of the electrolyte to the surface results in the fact that the corrosion potential values depend on the chromium content at the surface. In contrast, the tightness of surface layers is high enough to generate additional corrosion cells between layers and the substrate material resulting in increased intensity of corrosion processes, which in turn results in a high value of corrosion current densities. A simultaneous occurrence of both elements which control corrosion processes continues during potentiodynamic measurements up to such a potential value when macroscopic breakdown of layers and a dramatic increase in current density occurs – electrolysis.

4. Conclusions

Within the framework of the discussed research, the analysis of the influence of laser radiation fluence generating colour layers was carried out on the following three different types of stainless steel: AISI 304, AISI 316 and AISI 321 on the physical and chemical structure and the corrosion resistance of these layers. The surface analysis of layers with the use of SEM scanning microscope with EDX adapter was carried out. Analyzing the obtained results it was found that depending on the value of accumulated fluence the contents of particular alloy elements (Fe, Cr, Ni, O) in subsurface layers undergo changes. Migration of chromium from the surface to the subsurface layers was also found, which significantly formed corrosion resistance and the scope of migration depended on laser radiation fluence.
The change in the corrosion resistance of particular colourful layers on each type of steel in a fluence function of laser radiation was determined in electrochemical experiments with the use of a potentiodynamic method. The results of comprehensively compiled research methods make it possible not only to determine the influence of the value of laser radiation fluence on corrosion resistance of

the layers produced under its influence, but also to determine the mechanisms of corrosion occurrences there. By analyzing the obtained results a discussion about possibly occurring phenomena and their mechanisms was generated. It was also found that there is a possibility of producing layers with very high corrosion resistance and the scope of production parameters of such layers was specified.

References

[1] Laakso P., et al., *12th Conference on Laser Processing of Materials in the Nordic Countries (NOLAMP)*, Copenhagen, August 24–26. 2009
[2] Westin, Elin M., et al. "Effect of laser color marking on the corrosion performance of stainless steel.", Presented at ICALEO 2009
[3] O'Hana S., "Jewellery design using lasers: an artist's view.", Laser User 47 (2007) 32–132
[4] Antończak, A.J., et al., "The influence of process parameters on the laser-induced coloring of titanium.", Applied Physics A 115.3 (2014): 1003-1013
[5] Li, Z. L., et al., "Analysis of oxide formation induced by UV laser coloration of stainless steel.", Applied Surface Science 256.5 (2009): 1582-1588
[6] S. Krishnan, J. Dumbre, S. Bhatt, Esther T. Akinlabi, and R. Ramalingam, "Effect of Crystallographic Orientation on the Pitting Corrosion Resistance of Laser Surface Melted AISI 304L Austenitic Stainless Steel.", International Scholarly and Scientific Research & Innovation 7(4) 2013
[7] T. Dikova, M. Balcheva, N. Panova, *Machines, Technologies, Materials.* ISSN 1313-0226. ISSUE 9/2013
[8] N-EN ISO 17475, "Corrosion of metals and alloys – Electrochemical test methods – Guidelines for conducting potentiostatic and potentiodynamic polarization measurements." (ISO 17475:2005/Cor 1:2006)
[9] Wikipedia, „Stainless steel," 21 1 2015. [Online]. Available: http://en.wikipedia.org/w/index.php?title=Stainless_steel&oldid=643496764
[10] J. Goldstein, D. Newbury, D. Joy, C. Lyman, P. Echlin, E. Lifshin, L. Sawyer i J. Michael, *Springer* US, 2003
[11] Z. Li, H. Zheng, K. Teh, Y. Liu, G. Lim, H. Seng i N. Yakovlev, *Applied Surface Science*, 256, (5) 1582-1588, 2009
[12] N.E. Hakiki, M.F. Montemor, M.G.S. Ferreira, M. da Cunha Belo, *Corrosion Science* 42 (2000) 687-702

Acknowledgements
This research was supported by Wroclaw Research Centre EIT+ under the project "The Application of Nanotechnology in Advanced Materials" – NanoMat (POIG.01.01.02-02-002/08) financed by the European Regional Development Fund (Innovative Economy Operational Programme, 1.1.2).

Applying of non-toxic oxide alloys and hybrid polianiline compounds as anticorrosive pigments in organic epoxy coatings

W Szymański, A Halama[1] and J Madaliński
Electrotechnical Institute Division of Electrotechnology and Materials Science M. Sklodowskiej-Curie 55/61 50-369 Wroclaw, POLAND

E-mail: halama@iel.wroc.pl

Abstract. The objective of this work was to study inorganic oxide pigments as well as polyaniline heptamolybdes anticorrosive efficiency in epoxy coating. Antycorrosion resistance of modified coatings was examined by accelerated corrosion test in comparison to coatings of the suitable commercial epoxy paint. The carried out investigations shoved much bigrs anticorrosion performance of coatings modified with elaborated, new pigments.

1. Introduction

Growing demands of health and environment protection, drastically restricted application in organic coatings, highly effective but toxic chromate and lead active pigments. Applying of harmful for water organism zinc phosphate was also strongly limited. Thus, in the world industry of an anticorrosive paints raised demand of new, non-toxic and efficient active pigments. For the last twenty years many research attempts were devoted to use of conducting polymers (CP) as anticorrosion pigment in the organic coatings. Among them, thanks to good stability, low cost, low toxicity and good electrochemical properties, polyaniline(PANI) gained the highest importance. However, foregoing research works entirely applied basic form of PANI (emeraldine EB) or its salts with mineral acids (emeraldine ES). Presumably, because of applied toilsome method of their producing, the possibility of appliance of hybrid organic-inorganic PANI compounds hasn't been studied yet. The objective of this work was to study in organic oxide pigment and hybrid PANI compounds anticorrosion action in epoxy coating.

2. Inorganic anticorrosive pigments

Ten, active inorganic pigments with different stoichiometric composition have been obtained by high temperature synthesis. As a result of thermal decomposition of selected initial reagents, such as: zinc phosphate, ammonium heptamolybdate, boric acid, calcium and sodium carbonates, raised oxides entering the composition of well know anticorrosion pigments and possessing the ability to form glassy phase by their alloys.
Molar portions of particular oxides in received alloys are presented in table 1.

[1] Agnieszka Halama PhD

Table 1. Molar portions of oxides in received alloys.

Pigment	1	2	3	4	5	6	7	8	9	10
					[% mol]					
ZnO	22.68	22.22	21.84	21.48	20.97	22.84	22.97	23.11	23.26	23.44
P_2O_5	7.56	7.41	7.28	7.16	6.99	7.61	7.66	7.70	7.75	7.81
MoO_3	12.49	10.92	9.67	8.45	6.69	12.57	12.64	12.72	12.80	12.90
CaO	16.91	19.92	22.34	24.68	28.04	18.44	20.02	21.58	23.21	24.84
B_2O_3	25.47	24.95	24.53	24.13	23.55	25.65	25.79	25.95	26.12	26.32
Na_2O	14.89	14.58	14.34	14.10	13.76	12.90	10.92	8.93	6.86	4.70

After theirs grinding to grain diameter <10 μm, an appropriate solubility of obtained pigments was proved in agreement to suitable standard. The coefficients of inhibitive efficiency of particular pigments were determined by potentiodynamic and impedance spectroscopy methods in 3.5 mas % NaCl solution, and are presented in figure 1.

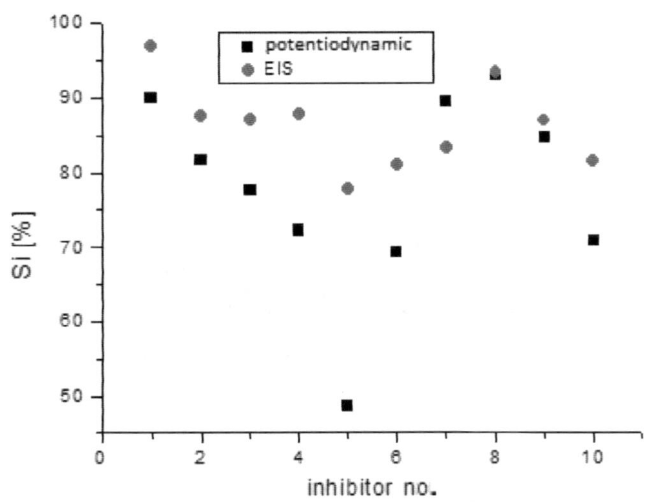

Figure 1. The coefficients of anticorrosive efficiency of inorganic pigments.

The highest efficiency was showed by pigments no. 1 and no. 8 according to table 1. In amount of 1 mas % grinded pigments were emulgated in a resin of epoxy paint not containing any other additions. The test plates covered with modified coatings were tested in wet SO_2 atmosphere in comparison to coating of suitable commercial epoxy paint pigmented with zinc phosphate (0 plate). The plates after 20 daily cycles of testing were presented in figure 2.

The best anticorrosion resistance showed coating modified with pigments no. 1, 4, 6 and 10, which remain almost unchanged after testing, whereas reference coating showed scratch corrosion in the range of 2 mm and about 3% of a roasted surface.

3. Polyaniline pigments

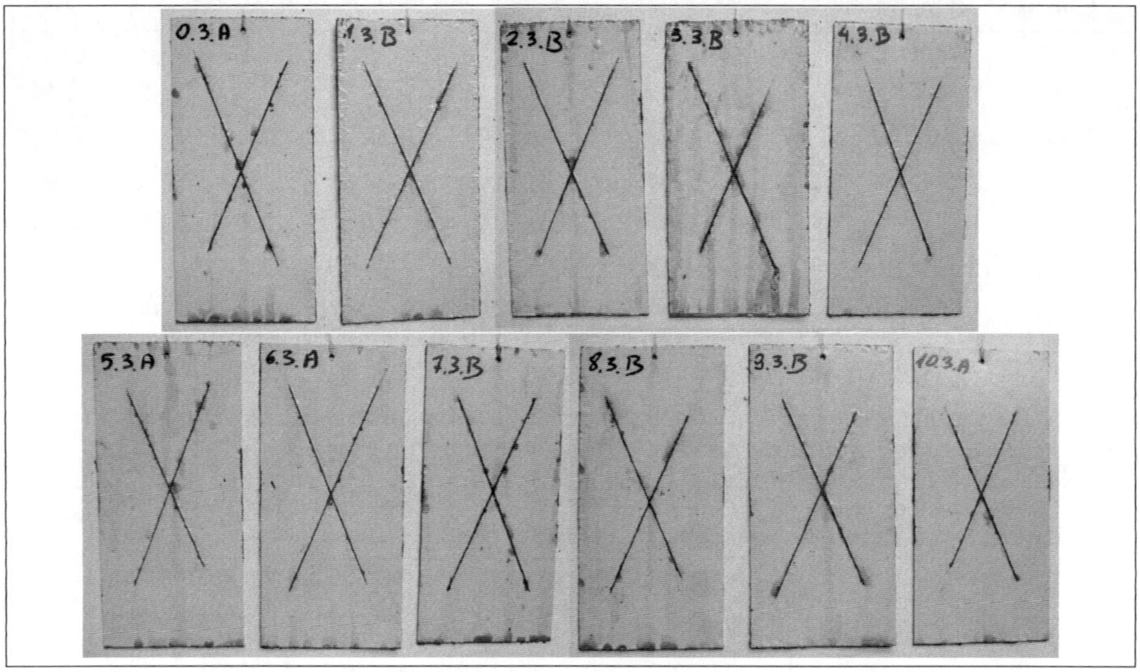

Figure 2. Test plates after stressing in wet SO_2 corrosion chamber.

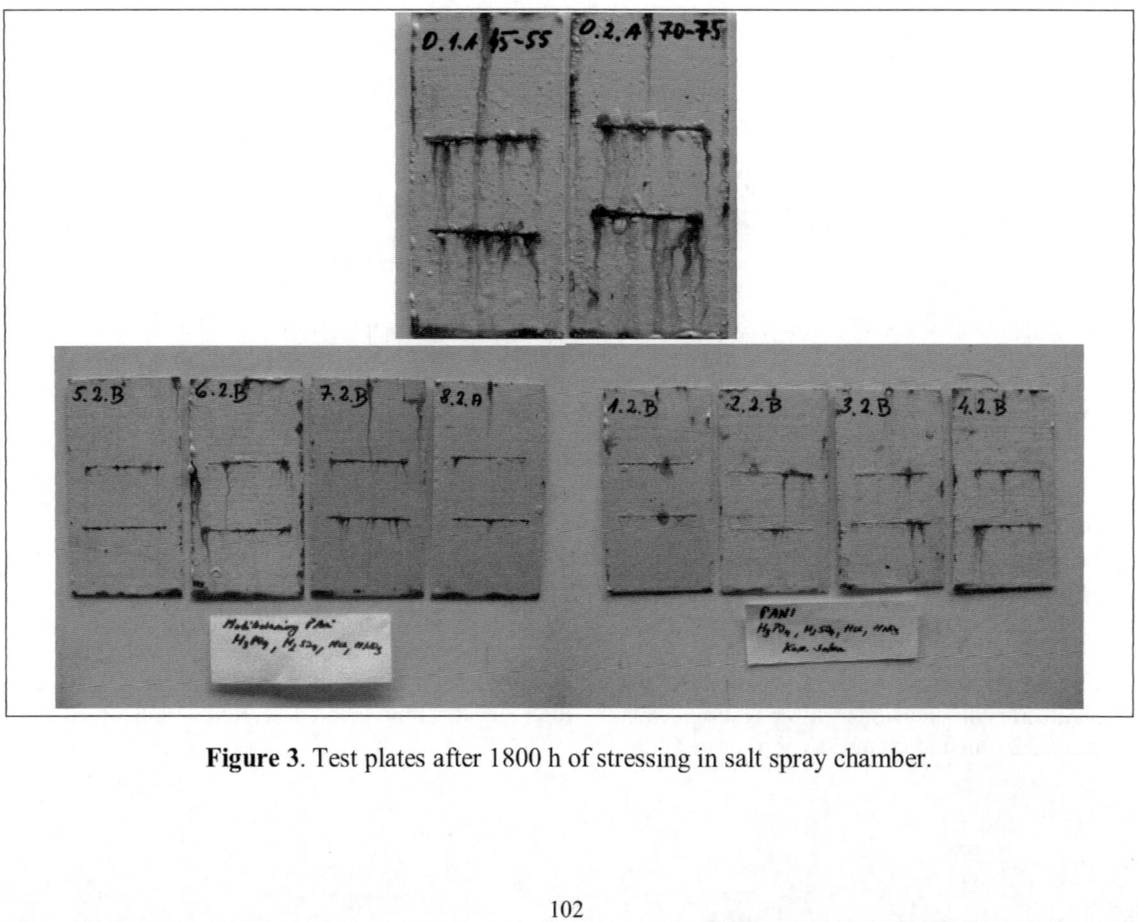

Figure 3. Test plates after 1800 h of stressing in salt spray chamber.

The polyaniline heptamolybdates as anticorrosive pigments. Applying a new simple method of synthesis, four forms of PANI heptamolybdates were received in solutions of phosphoric, sulphuric, hydrochloric and nitric acids. Depending on used acid, an obtained compounds showed different FTIR spectra and characteristics of thermal analysis curves. These compounds as well as PANI salts obtained in solutions of the same acids, were applied in an epoxy resin with concentration of 0.5 mas %.

The modified paints were deposited on the test plates and subjected to accelerated corrosion testing in neutral salt spray together with reference coating of appropriate commercial epoxy paint.

The plates after 1800 h of testing are presented in figure 3.

The reference coatings of commercial epoxy paint showed distinct blistering and very extensive scratch corrosion in the range 10-15 mm. The coating which was modified with PANI salts showed scratch corrosion in the range of 3-4 mm and much lower level of blistering, whereas the coatings containing PANI heptamolybdates received in solution of H3PO4 and HNO3 acids (plates 5.2.B and 8.2.A) except of scratch roasting, were not distinctly damaged after testing.

4. Conclusions

The preliminary estimated cost of producing and relatively low level of application in the coating with satisfactory anticorrosion performance makes the most effective from tested pigments, competitive to commonly and currently used zinc phosphate. The little zinc concentration in inorganic pigments, together with relative, low level of their concentration in organic coating, cause their practical applying quite allowable. The use of presented pigments in practice, may contribute to bridge the technological gap ensuing in the world industry of anticorrosive paints, after limitation in using pigments harmful for health and an environment.

5. References

[1] Marti M 2012 *Prog. Org. Coat.73* 321
[2] Zubielewicz M 2014 *Ochrona przed Korozją* **9** 347
[3] Sorensen P A, Kill S and Dam-Johansen K 2009 *J. Coat. Technol. Res.* **6** 135
[4] Riaz U, Nwoha C and Ashraf S M 2014 *Prog. Org. Coat.* **77** 743
[5] Rohweder M, Minh Duc L and Michalik A 2007 *Electrochim. Acta* **53** 1300
[6] Armelin E 2007 *Prog. Org. Coat.* **59** 46
[7] Armelin E Aleman C Iribarren J I 2009 *Prog. Org. Coat.* **65** 88
[8] Zarras P *Rad. Phys. Chem.* 2002 **68** 43
[9] Tallman D E 2002 *J.Solid State Electrochem.* **6** 73
[10] Spinks G M 2002 *J.Solid State Electrochem.* **6** 85
[11] Epstein A J Handbook of Conducting Polymers, 3rd ed, 2007, New York CRC Press, 15
[12] Gomez-Romero P, Casan-Pastor N Lira-Cantu M 1997 *Solid State Ionic* **101-103** 875
[13] D.E.Shengxian et all Coating composition and method of making and using them. US Pat Appl. Pub. No US/2006/034339A1, 2006

Acknowledgments. The research was supported by Wroclaw Research Centre EIT+ within the project "The Application of Nanotechnology in Advanced Materials" - NanoMat (POIG.01.01.02-02-002/08) financed by the European Regional Development Fund (Innovative Economy Operational Programme, 1.1.2).

IOP Conf. Series: Materials Science and Engineering **113** (2016) 012019 doi:10.1088/1757-899X/113/1/012019

Analysis of electric current flow through the HTc multilayered superconductors

J Sosnowski
Electrotechnical Institute
Pożaryskiego 28, 04-703 Warsaw, Poland

E-mail: j.sosnowski@iel.waw.pl

Abstract. Issue of the flow of the transport current through multilayered high-temperature superconductors is considered, depending on the direction of the electric current towards the surface of the superconducting CuO_2 layers. For configuration of the current flow inside of the layers and for perpendicular magnetic field, it will be considered the current limitations connected with interaction of pancake type vortices with nano-sized defects, created among other during fast neutrons irradiation. So it makes this issue associated with work of nuclear energy devices, like tokamak ITER, LHC and actually developed accelerator Nuclotron-NICA, as well as cryocables. Phenomenological analysis of the pinning potential barrier formation will be in the paper given, which determines critical current flow inside the plane. Comparison of theoretical model with experimental data will be presented too as well as influence of fast neutrons irradiation dose on critical current calculated. For current direction perpendicular to superconducting planes the current-voltage characteristics are calculated basing on model assuming formation of long intrinsic Josephson's junctions in layered HTc superconductors.

1. Introduction

High temperature oxide superconductors are characterised by specific crystal structure in which copper-oxide layers are alternated with buffer layers, providing the electric charge to them. Electric properties of such subtle two dimensional structure of CuO_2planes and of CuO one dimensional chains, are very sensitive to the disturbances caused for instance by fast neutrons irradiation. Influence of nano-defects on current-voltage characteristics of HTc superconductors for intra-layered current is investigated in first part of this paper, while in second it will be given analysis of interlayered currents. In Fig. 1 is shown the creation of nano-sized defects as the result of the fast neutrons irradiation, in the form of vacancies (□), as well as inter-site atoms (0), raised in secondary process of knocking them out by initially irradiated atoms.

IOP Conf. Series: Materials Science and Engineering **113** (2016) 012019 doi:10.1088/1757-899X/113/1/012019

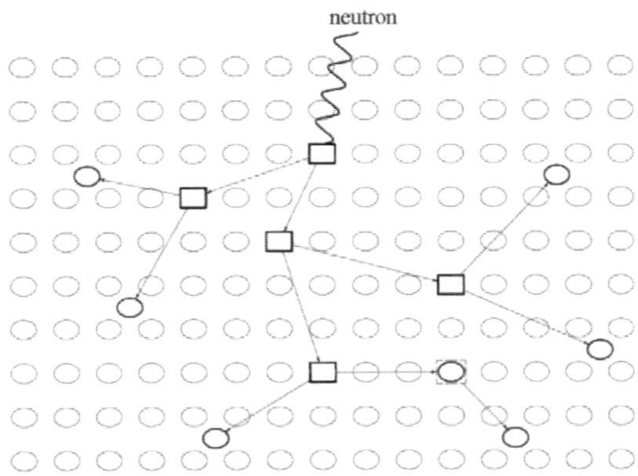

Figure 1. Scheme of the creation of nano-sized defects during the fast neutrons irradiation process

2. Current-voltage characteristics analysis of multi-layered HTc superconductors

Important task from the point of view of applications HTc superconductors in power devices as cryocables, investigated in the present paper, is issue in which way structure of HTc multi-layered superconductors influences the process of current flow through these materials. The inside-plane critical current has been determined by interaction of magnetic flat pancake type vortices with nano-sized defects, while multilayered structure of HTc superconductors influences this interaction because pancake vortices from neighbouring layers attract, while in the same layer repulse. Perpendicular to layers critical current has been described by intrinsic Josephson's effects. Analysis of energy [1] of captured on nano-sized pinning centre pancake vortex allows to determine then potential barrier which after taking into account the elasticity energy of vortex lattice brings the final form:

$$\Delta U(i) = \alpha_{\downarrow} e(^{\uparrow}2(\nu(1-i^{\uparrow}2)-2)\nu(1-i^{\uparrow}2)+\Delta U_{\downarrow}L(i) \quad (1)$$

In Eq. 1 Hc is thermodynamic critical magnetic field, l thickness of superconducting layer, d width of pinning center, ξ coherence length, parameter αe describes elasticity shear module of vortex lattice, i=j/jC is reduced to critical jC, transport current density j. Expression on potential barrier for flux creep process ΔU allows to determine then current-voltage (I-V) characteristics.

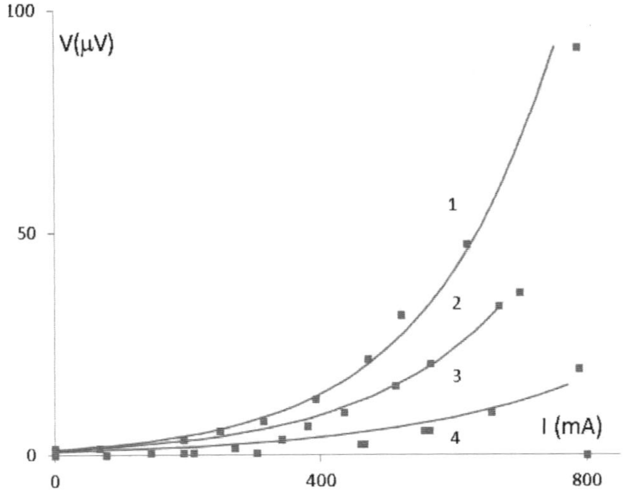

Figure 2. Fitting of theoretical model to experimental current – voltage characteristics for the HTc superconductor of composition BiSrCaCuO doped Pb at 77 K, in various magnetic fields: (1) B = 21.5 mT, (2) 16.3 mT, (3) 8.5 mT, (4) 0 T

Results of fitting theoretical model to experimental data measured on BiPbSrCaCuO sample is shown in Fig. 3 in the function of magnetic induction. Quite good agreement is observed here. Fig. 3 presents the results of calculations in this model of the influence of fast neutrons irradiation on the critical current of HTc superconductor. Initial increase of critical current is observed for pure sample, which is followed then by the gradual decrease of critical current, connected with damaging of superconductor structure.

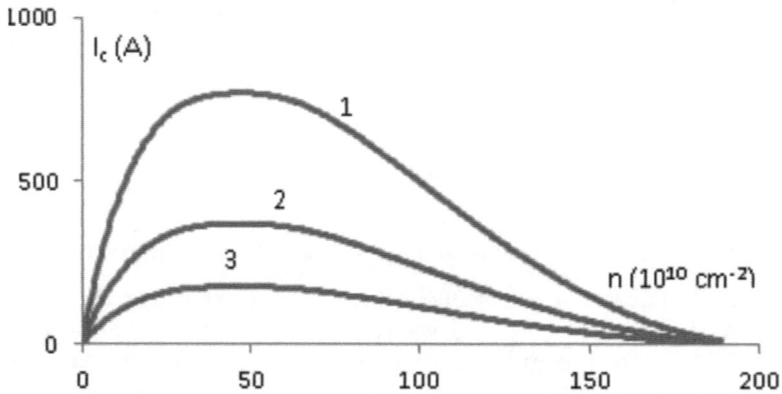

Figure. 3. Calculated critical current dependence on surface concentration (*n*) of fast neutrons irradiation in the function of magnetic induction: (1) $B = 0.5$ T, (2) $B = 1$ T, (3) $B = 3$ T for $T = 20$ K

Transport of perpendicular to the surface of multi-layered superconductor current has been treated in the paper basing on Josephson's effects analysis. It lead to sinusoidal relation joining the current flow through the junction and phase difference of wave function $\Delta\Phi$ between subsequent layers, in the function of magnetic flux Φ, reduced to magnetic flux quantum Φ_0. It has been noticed that in layered structure of HTc superconductor are formed long Josephson's junctions, for which phase difference $\Delta\Phi$ fulfils following relation, known as sine-Gordon equation, describing propagation of solitons:

$$\frac{\partial^2 \Delta\phi}{\partial t^2} - \frac{\partial^2 \Delta\phi}{\partial x^2} + \alpha \frac{\partial \Delta\phi}{\partial t} + \sin \Delta\phi = \frac{J}{J_c} \qquad (2)$$

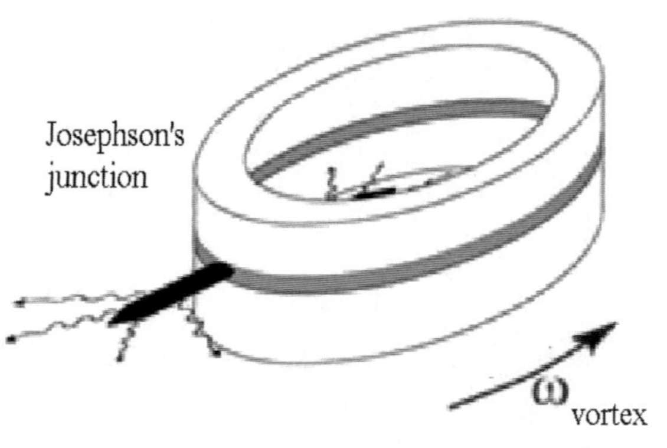

Figure 4. View of annular Josephson's junction with moving vortex emitting microwaves

Parameter $\alpha = \gamma\sqrt{\frac{\hbar}{2eCj_c}}$ in Eq. 2 denotes the normal part of total current connected with electrical conductivity γ and capacitance C of junction, leading to resistive, loss component of current in such long junction. The solution of Eq. 2 depends on considered geometry, which simplest case shown in Fig. 4, described by periodic boundary conditions, is annular junction of large diameter. Josephson's vortex moving in such junction treated as solition emits then radiation, which allows to use this junction as microwave generator. For such simple cylindrical geometry the solution of Eq. 2 has been found in perturbation procedure [2-4] in the form of travelling wave. Calculated according to this I-V characteristics are shown in Fig. 5 and indicate on significance of resistive component, described by parameter α for shape of these characteristics in long Josephson's junctions. It together with geometry of junction allows to receive appropriate range of emitted irradiation and sensitivity of device.

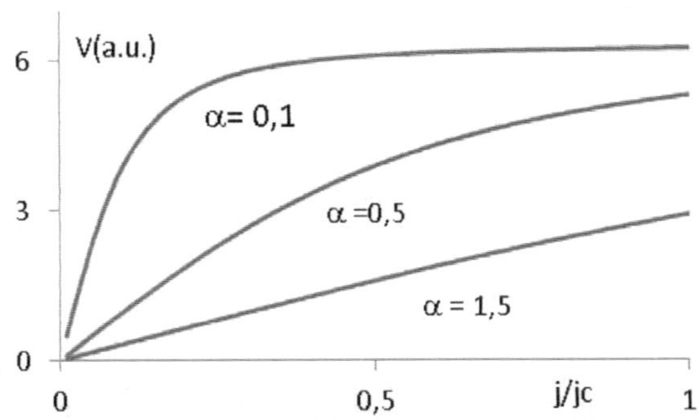

Fig. 5. Calculated influence of loss factor α on I-V characteristics of the long Josephson's junction

At this purpose there appear the projects to profile the geometrical shape of long Josephson's junctions and beside this annular shape also junctions like Eiffel's tower form and others are made. Appearance in such junctions of vortices allows to use them next in new type of electronic devices, which is called vortex electronics. Simplest case is using these junctions with vortex in digital logic devices, as computers. Appearance or lack of vortex can be treated as logic signal of the form yes or not, while great advantage of such solution is extremely short time of switching determined by Josephson's plasma frequency of the range picoseconds (10^{-12} s). Important would be too applications of these long junctions as emitters of radiofrequency waves, which idea is shown in Fig. 4.

3. Conclusions

In the paper is presented analysis of effects determining the process of flow electric current through multi-layered HTc oxide superconductors. Especial attention is directed to analysis of the inside-layers current and potential barrier formation, describing the capturing pancake vortices on nano-sized defects, arising especially in superconducting windings of the nuclear accelerators. The case of perpendicular to the superconducting layers current is investigated too basing on analysis of long Josephson's junction formation between superconducting layers. Phase shift of wave function between superconducting layers is described then by sine-Gordon equation. Possibility of application long Josephson's junctions in vortex electronics is described. Current-voltage characteristics have been calculated in both cases and fitted to experiment for pinning interaction, leading to good agreement.

4. References

[1] Sosnowski J, 2014 *Modern Phys. Letters B*, **28**, 1450132
[2] Pedersen N F, 1991 *IEEE Transactions on Magnetics*, **27**, 3329
[3] Marcus P M and Imry Y, 1980 *Sol. State Comm.* **33**, 345
[4] McLaughlin D W and Scott A C, 1978 *Phys. Rev. A* **18**, 1652

IOP Conf. Series: Materials Science and Engineering **113** (2016) 012020 doi:10.1088/1757-899X/113/1/012020

Acoustic emission during fracture of ceramic superconducting materials

L Woźny, A Kisiel and K Łysy
Wrocław University of Technology, Faculty of Electrical Engineering, 27 Wybrzeże Wyspiańskiego St., 50-370 Wrocław, Poland

E-mail: leszek.wozny@pwr.edu.pl, anna.kisiel@pwr.edu.pl

Abstract. In the ceramic materials acoustic emission (AE) is associated with a rapid elastic energy release due to the formation and expansion of cracks, which causes generation and propagation of the elastic wave. AE pulses measurement allows monitoring of internal stresses changes and the development of macro- and micro-cracks in ceramic materials, and that in turn allows us to evaluate the time to failure of the object. In presented work the acoustic signals generated during cracking of superconducting ceramics were recorded. Results obtained were compared with other ceramic materials tested the same way. An analysis of the signals was carried out. The characteristics of the AE before destruction of the sample were determined, that allow the assessment of the condition of the material during operation and its expected lifetime.

1. Introduction

Year of 1986 was a turning point for superconductivity, because in that year Johannes G. Bednorz and KA Müller discovered superconductivity in ceramic material La-Ba-Cu-O [1]. This discovery was specific, since the critical temperature T_c of this compound is around 30-35 K (approximately -238°C). This temperature is 10°C higher than the highest critical temperatures of conventional superconductors, which are metals, alloys and intermetallic compounds. Since that moment there have been many discoveries of high temperature superconductors based on copper oxide.

One of these is $YBa_2Cu_3O_x$ [2]. This is the first compound with reproducible properties and, what is important, with T_c=90-95 K, which is above the boiling point of liquid nitrogen (77 K). The possibility of cooling the new superconductors by liquid nitrogen is their unquestionable advantage because this coolant is much cheaper and more accessible than liquid helium, which is used for conventional superconductors cooling.

High-temperature superconducting ceramics in most cases containing copper and the copper-oxygen planes (existing in their structure) have a dominant influence on their properties. These compounds have black colour, are hard and brittle and are characterized by high porosity.

Superconducting ceramics are characterized by strongly irreversible magnetization curves thus they are type II superconductors. Their value of the upper magnetic critical field reaches is very high. However, the value of the lower magnetic critical field is relatively small, so that it is easy to induce the mixed state in these materials.

Superconducting ceramics is a brittle substance. It shows a lower real resistance compared with theoretical resistance. Therefore it is necessary to test its microstructure in order to ensure its reliability and safety. One of methods that can be applied to that study is the acoustic emission (AE) technique [3]. Its big advantage is fastness and non-destructive nature of the research. The phenomenon of acoustic emission allows monitoring changes in internal stresses and the development of macro- and micro-cracks in ceramic materials [4-5]. This in turn allows to evaluate the remaining time to the destruction of the object, etc. [3].

In the ceramic materials acoustic emission is associated with a rapid energy release caused by the formation and expansion of cracks, whereby there is a generation and propagation of the elastic wave. Energy can be released due to the different processes occurring in the material, both in submicroscopic and macroscopic scale. The sub-micron processes include, among others, a jump position adjacent atoms in the crystal lattice, while the macroscopic effect is a catastrophic destruction of the material.

In the presented paper it was carried out preliminary research on acoustic emission during fracture samples of ceramic superconductors $YBa_2Cu_3O_x$ under the influence of external mechanical action. For comparative purposes, similar tests were conducted for samples of electrical porcelain.

2. Acoustic emission technique

The measurement method consists in recording the AE pulses which are generated as a result of propagation micro damage or changes of stress inside material. Materials that can be studied by acoustic emission are: concrete, reinforced concrete, ceramics, glass, metals, composites, laminates, etc. Each material has a primary or acquired during the operation heterogeneous distribution of elastic energy (residual stresses) and a certain level of defects in the structure, and even micro- and macro damages [2]. Any external factor (change of temperature, applied force) can change this state and release portions of energy in the form of elastic waves. These waves can be registered as AE pulses.

The block diagram of setup for recording acoustic emission is shown in figure 1.

Figure 1. A block diagram for the AE measurement. D - piezoelectric detector, PM - preamplifier, FG – high pass filter, WG - main amplifier, A - discriminator threshold, L - counter, C/A - digital to analog converter, R - recorder.

Piezoelectric detector is the most important part of the apparatus. Its task is to transform ultrasonic wave energy to the energy of the electrical signal [5]. This change is connected with the piezoelectric effect during mechanical deformation. Between the walls of the piezoelectric crystal transducer it is formed the potential difference proportional to the force inducing strain [3]. The transmitter housing is the most common shape of a cylinder and is made of high quality stainless steel to eliminate electromagnetic interference.

The detector is acoustically coupled with the test object and electrically connected with the preamplifier. A preamplifier with a low noise level provides preamplifier and matching between the transducer and the transmission cable. Then the signals go to a high-pass filter, where noises are filtered through cutting the bandwidth from the bottom. Next the signals are amplified by the main amplifier. AE signal reaches the threshold discriminator, whose task is to pass signals above the fixed amplitude. Pulses with an amplitude exceeding the lower threshold are counted in the counter and transmitted by the D/A converter to the recorder. The result of measurement is the number of counted pulses of acoustic emission whose amplitude is between the lower and upper threshold of amplitude discriminator.

3. Cracks in ceramics

Crack is macroscopic gap located in the material. The propagation of cracks is the main source of EA in ceramics [2]. Cracks and their development cause that the theoretical strength is different than the real strength. This is due to the presence of defects in the material structure. Defects can be divided into two types.

Defects of first kind include all kinds of stress concentrators in the form of sharp gaps or notches of any shape - they are hence the defects of a geometrical nature, not related to the structure and construction material.

Defects of the second kind are stress concentrators in the form of dislocations, voids deployed along the grain boundaries, inclusions of foreign material (eg. the coal in metals) causing contact stresses and all other defects in the internal structure of the material.

There are described 3 crack deformation modes that are distinguished because of the way to get around the edges of the gap: mode I - tearing, mode II - transverse shear, mode III - longitudinal shear. The crack is in the form of shear, if its surfaces diverge in the direction perpendicular to its front. The transverse shear occurs when the crack surfaces slide against each other toward the front. The longitudinal shear occurs when the crack surfaces slide past each other in a direction parallel to the front of the slot.

Analysis of the body with defects in the structure was described by Griffith theory [6]. In the stress state, the dependence between the free energy and the length of the slot has maximum value, which constitutes a barrier for the spread of cracks. For this reason, in the fixed state of stress it can be developed only these defects, which lengths are greater than critical length that corresponds to the maximum value of the free energy.

4. The measurement results

Measuring setup used for measurement of AE consisted of a source of external stress (steel jaws), that evoked cracking in the sample by its compression. The piezoelectric transducer was pressed to the source of stress. It was connected with other parts of the measurement system.

The tested materials were electrotechnical porcelain and ceramic $YBa_2Cu_3O_x$ superconductors.

Figures 2 and 3 present dependences of the AE rate on time during slowly destroying by crushing for porcelain and superconducting ceramics respectively.

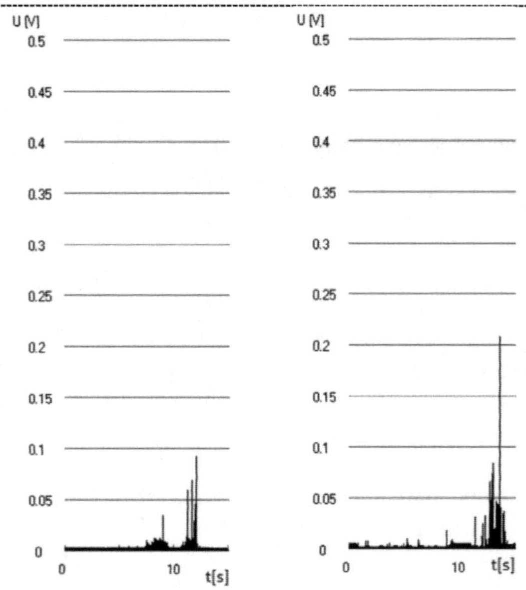

Figure 2. The amplitude of the AE for electrotechnical porcelain during cracking.

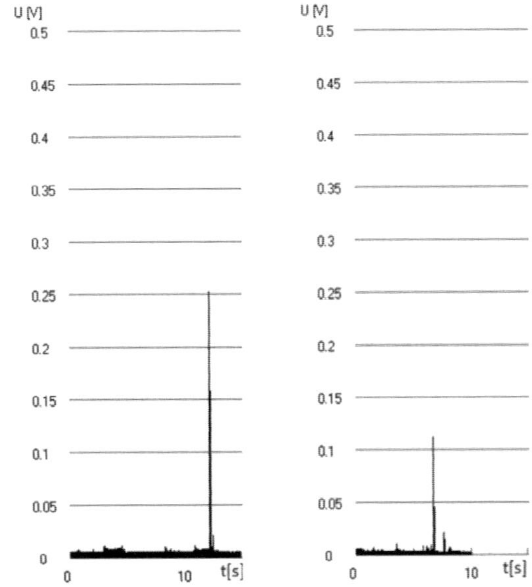

Figure 3. The amplitude of the AE for YBCO superconducting ceramics during cracking.

Waveforms for both samples are presented in figures 4 and 5. Figures 6 and 7 show the detailed frequency characteristics for them.

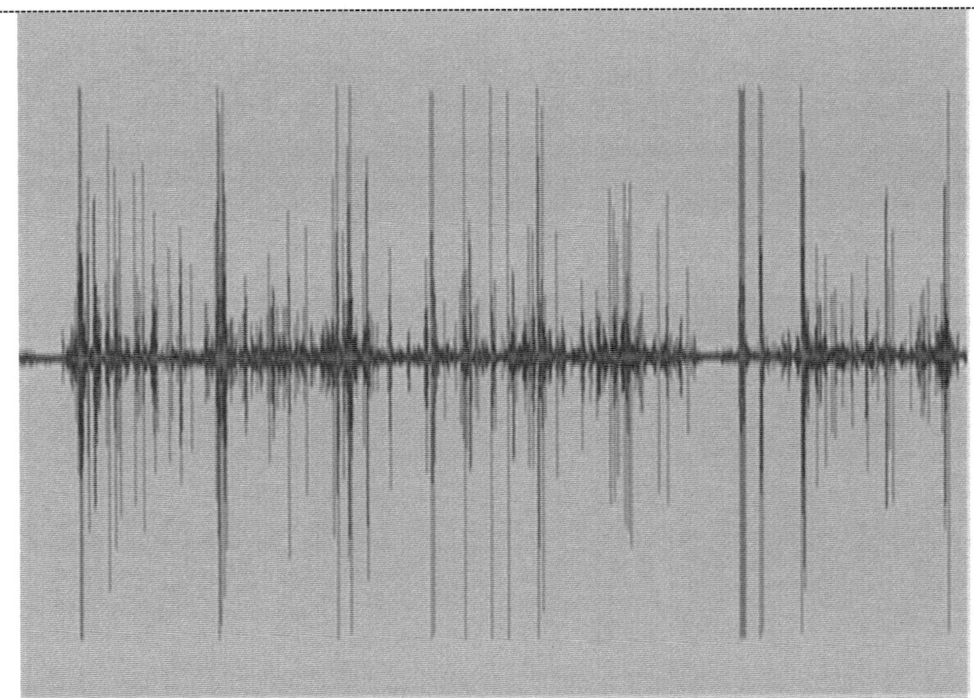

Figure 4. Oscilloscope image of AE pulses during cracking of electrotechnical porcelain.

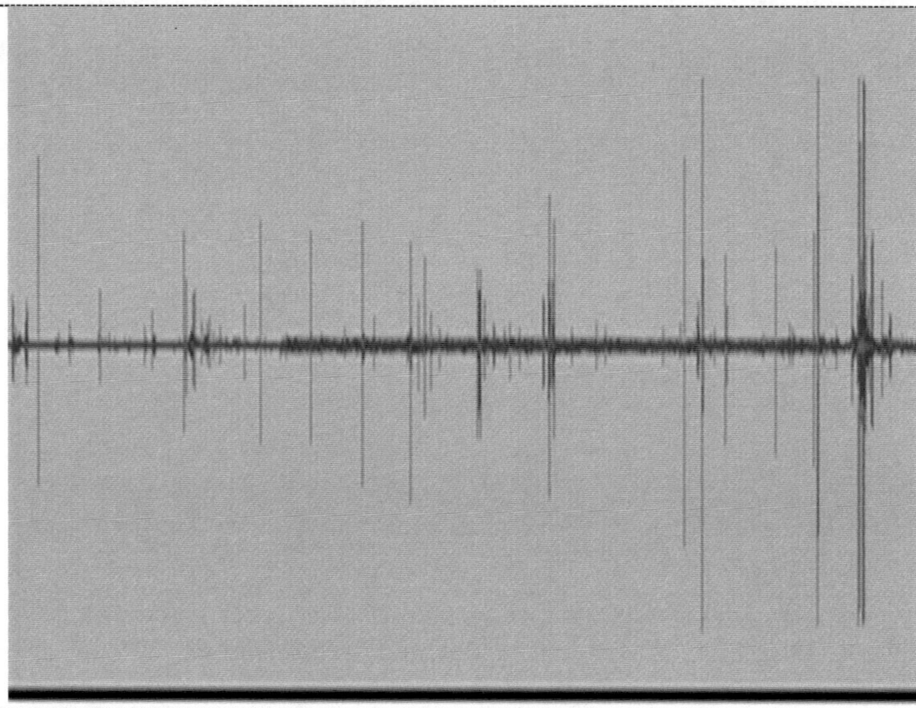

Figure 5. Oscilloscope image of AE pulses during cracking of ceramic superconducting.

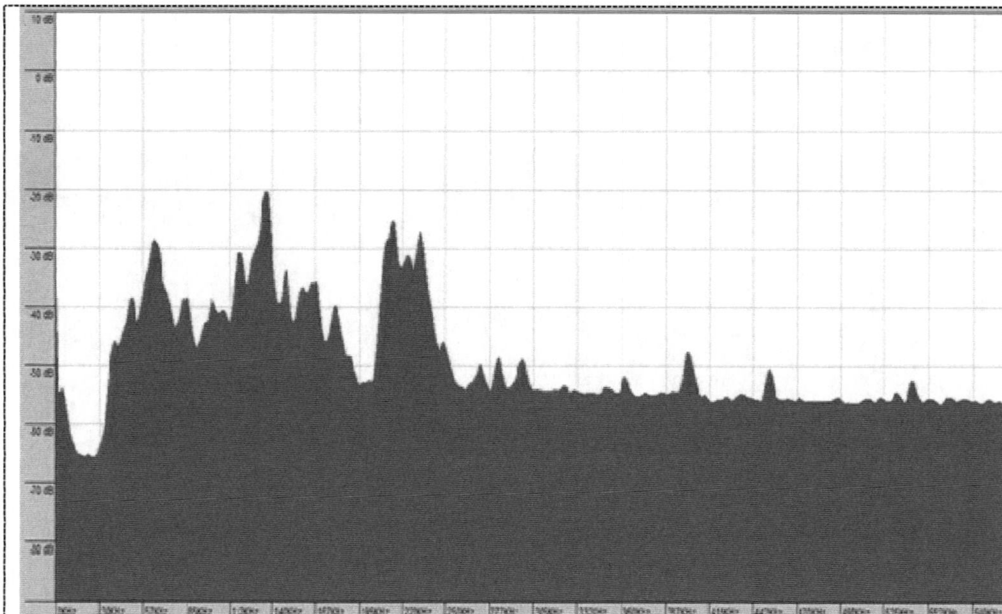

Figure 6. The frequency characteristic for electrotechnical porcelain during cracking.

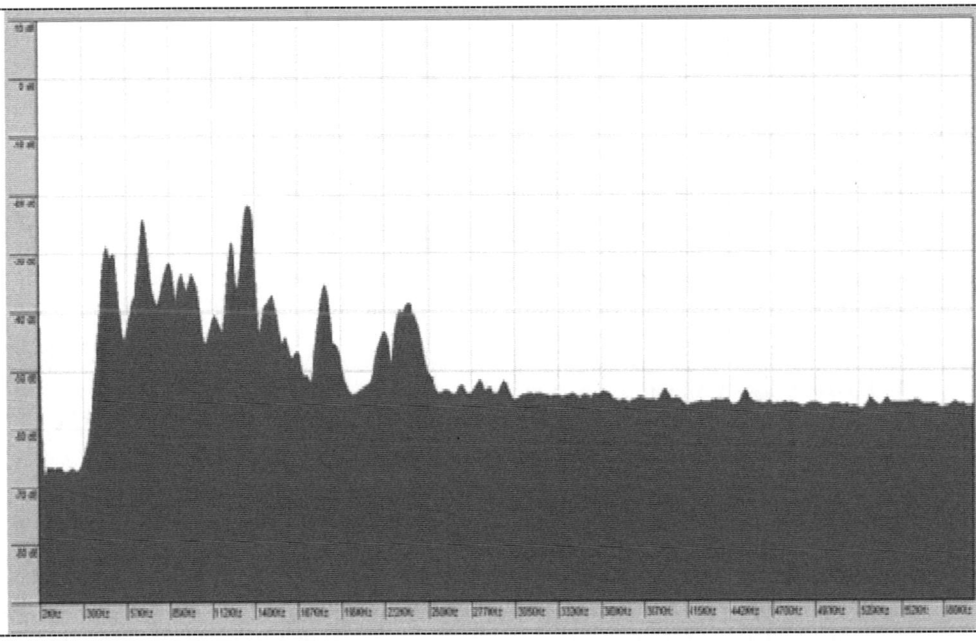

Figure 7. The frequency characteristic for YBCO ceramics during cracking.

5. Conclusions

On the basis of the test results the following conclusions can be drawn:

- Acoustic emission during fracture of superconducting ceramics is less intense than in electrotechnical porcelain. Both the amplitude and pulse sum AE for YBaCuO is significantly lower.
- The reason for this could be the difference in defect structure of materials, for example different porosity.
- The dominant frequency recorded during cracking superconducting ceramics and porcelain materials are very similar (65-80 kHz and 130-140 kHz).

An extension of the work should be to examine the AE during ceramic superconducting cracking at low temperatures (in the superconducting state).

6. References

[1] Bednorz J G and Müller K A, 1986 *Z. Phys. B Cond. Mat.* **64** 189
[2] Wu M K, Ashburn J R, Torng C J, Hor P H, Meng R L, Gao L, Huang Z J, Wang Y Q and Chu C W 1987 *Phys. Rev. Lett.* **58** 908
[3] Roberts T M and Talebzadeh M 2003 *J. Constr. Steel Res.* **59** 695
[4] Zeng J, Yong H D and Zhou Y H 2011 *J. Appl. Phys.* **109** 093920
[5] Gao Z, Zheng Z and Li X, 2015 *Physica C* **519** 5
[6] Griffith A A 1920 *Philosophical Transactions, Series A* **221** 163

Acknowledgments
This work was funded by the statutory research.

The interaction between the permanent magnet and ceramic superconductor with organic filler

L Woźny, A Kisiel and A Garbera
Wrocław University of Technology, Faculty of Electrical Engineering, 27 Wybrzeże Wyspiańskiego St., 50-370 Wrocław, Poland

E-mail: leszek.wozny@pwr.edu.pl, anna.kisiel@pwr.edu.pl

Abstract. The aim of the study was to examine the phenomenon of magnetic levitation for YBaCuO superconducting samples in pure form and with epoxy resin content of 40%. Samples of superconductors were prepared by the standard reaction in the solid state. The forces of interaction between the superconductor and neodymium permanent magnet were measured. Samples with epoxy resin fillers had significantly smaller levitation force than the sample of the sintered superconductors. This is due to a much lower content of pure superconducting material in the sample volume (about 60% of the YBaCuO). However, the obvious advantage of such samples is the possibility of preparation superconductors with complicated shapes, eg. for use in a superconducting bearings or other devices.

1. Introduction

Magnetic levitation phenomenon is directly related to the Meissner effect, that is, repulsion the induction of the external magnetic field from the inside of superconducting material [1-2]. If the source of the magnetic field is a permanent magnet, it occurs repulsion effect between magnet and superconductor. From the application point of view, it is important to determine the magnetic levitation force that can be really achieved in practice.

The phenomenon of magnetic levitation, due to its properties can be used for

- high-speed bearings - with virtually no losses due to friction [3],
- shock absorbers for precision equipment - levitating magnet dampens vibrations [4],
- levitating railway construction [5-6].

Levitation force depends crucially on the quality and type of superconducting samples, their composition, mass and properties of permanent magnets. For practical applications of this phenomenon it becomes increasingly important to obtain the superconducting materials of any shape. This can be achieved by adding to the powdered superconductor adhesive material. Therefore, the aim of this study was to determine the levitation force values for samples of superconductors in pure form and with the addition of epoxy resin.

2. Sample preparation

The well-known $YBa_2Cu_3O_x$ high-temperature ceramic superconductor has been used in work. Its synthesis was performed by standard solid state reaction at 950°C in an air atmosphere from Y_2O_3, $BaCO_3$ and CuO reagents.

Material obtained after grinding has been compressed and subjected to further annealing according to the following scheme

- temperature rise time: from 220 to 945°C, 4 h,
- the heating time at 945°C: 3 h,
- natural cooling to a temperature of 750°C,
- time annealing at 750°C: 1 hour,
- natural cooling process to room temperature.

The prepared material was thoroughly grinded and divided into two parts. The first portion was pressed, sintered at about 950°C and cooled in accordance with the previous procedure. The remaining powder was mixed with the epoxy resin (the content of YBaCuO - about 60% by volume).

3. The test results

Figure 1 shows a photograph of the magnets stably levitating over the YBaCuO sample in a superconducting state (77 K).

Figure 1. Permanent magnets levitating above the superconductor sample.

For measurements they were used type N42 neodymium magnets due to their very good parameters:

- remanence B_r = 13 kG,
- energy density BH = 330 kJ/m^3,
- good ratio of the weight of both the induction value B_r as well as energy density BH.

The levitation forces were measured by self-made setup. The measurement principle was based on the measurement of the mass of weights balancing repulsive force between the permanent magnet attached to one arm of the balance-pan and a superconducting material located below the magnet in the open cryostat. Repulsive force was measured for different values of the distance between the magnet and superconductor regulated by a micrometer.

Levitation force values depending on the distance between the superconductor and the permanent magnet for the sample superconductor without the epoxy resin is shown in figure 2 and for the sample with the addition of epoxy resin - in figure 3. Figure 4 shows a comparison of the levitation for both types of samples and for different sizes of magnets.

IOP Conf. Series: Materials Science and Engineering **113** (2016) 012021 doi:10.1088/1757-899X/113/1/012021

As would be expected, the force of interaction between a superconductor and a magnet decreases rapidly with increasing distance between them. This force is greater for the larger size of the interacting materials (figure 4). A similar effect can be observed by comparing the levitation force for pure samples of superconductors and with the addition of epoxy resin (figure 2, 3 and 4). In this latter case, the repulsive forces are much smaller. This is due to lower effective volume of the superconducting material in the samples with addition of the resin.

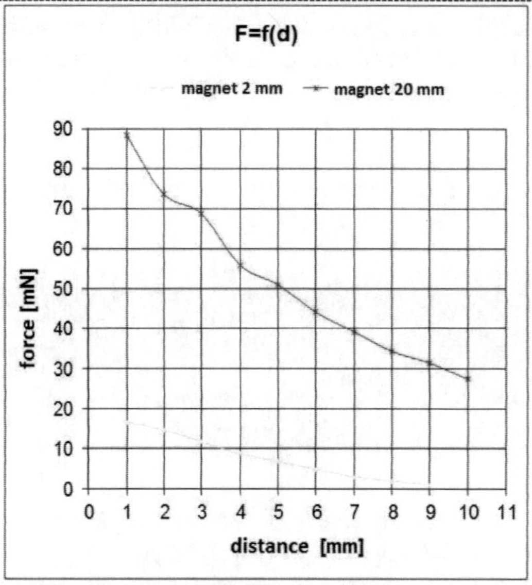

Figure 2. Dependence of levitation force on the distance for sample without epoxy resin.

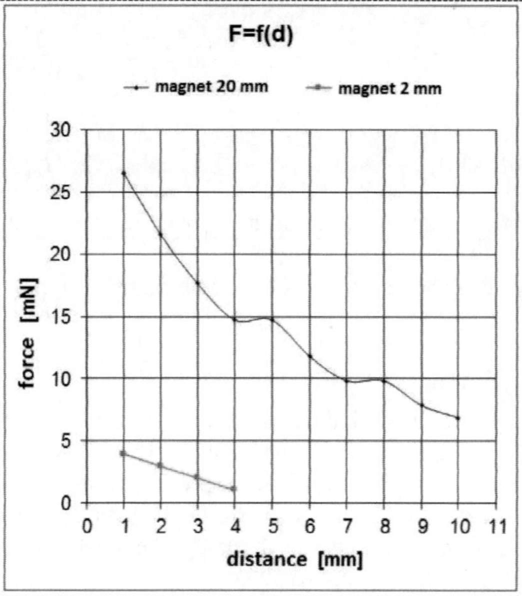

Figure 3. Dependence of levitation force on the distance from sample with epoxy resin.

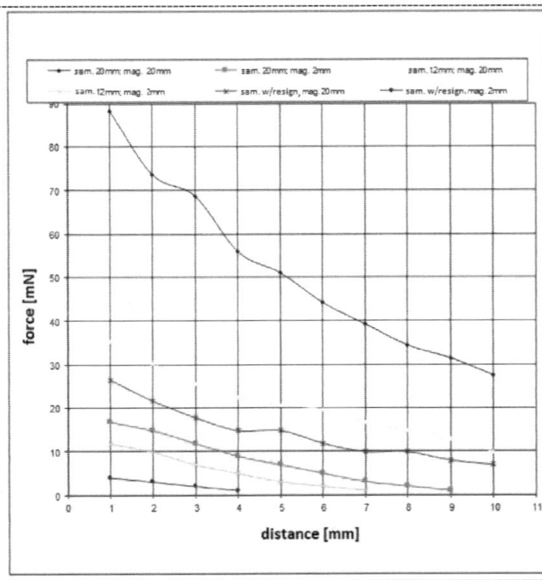

Figure 4. Levitation force depending on the type of samples and size of the magnets.

4. Conclusions

On the basis of the measurements it can be stated that:

- The results obtained in the work allowed to determine the levitation force depending on the type and weight of the samples and used magnets.
- The largest levitation forces were measured for the pure samples (without addition of epoxy resin) having the largest diameter and for magnets of the maximum dimensions.
- Samples with the addition of epoxy resin showed a reduced levitation force compared to the pure samples of the same size by the use the same magnets.

5. References

[1] Kordyuk A A 1998 *J. Appl. Phys.* **83** 601
[2] Okano M, Iwamoto T, Furuse M, Fuchino S and Ishii I 2006 *J. Phys. Conf. Ser.* **87** 999
[3] Hull J R, Mulcahy T M, Uherka K L and Abboud R G 1995 *IEEE Trans. Appl. Supercond.* **5** 626
[4] Teshima H, Tanaka M, Miyamoto K, Nohguchi K, and Hinata K 1997 *Physica C Supercond.* **274** 17
[5] Fujimoto H 1998 *J. Appl. Phys.* **50** 16
[6] Fujimoto H, Kamijo H, Higuchi T, Nakamura Y, Nagashima K, Murakami M and Sang-Im Yoo 1999 *JOM* **9** 301

Acknowledgments
This work was funded by the statutory research.

Direct waste heat recovery via thermoelectric materials – chosen issues of the thermodynamic description

Piotr Kolasiński[1] and Ewa Kolasińska[2]
[1]Wrocław University of Technology, Department of Thermodynamics, Theory of Machines and Thermal Systems, Wybrzeże Wyspiańskiego 27, 50-370, Wrocław, Poland
[2]Electrotechnical Institute, Division of Electrotechnology and Materials Science, M. Skłodowskiej-Curie 55/61, 50-369, Wrocław, Poland

E-mail: [1]piotr.kolasinski@pwr.edu.pl, [2]kolasinska@iel.wroc.pl

Abstract. The effective waste heat recovery is one of the present-day challenges in the industry and power engineering. The energy systems dedicated for waste heat conversion into electricity are usually characterized by low efficiency and are complicated in the design. The possibility of waste heat recovery via thermoelectric materials may be an interesting alternative to the currently used technologies. In particular, due to their material characteristics, conducting polymers may be competitive when compared with the power machinery and equipment. These materials can be used in a wide range of the geometries e.g. the bulk products, thin films, pristine form or composites and the others. In this article, the authors present selected issues related to the mathematical and thermodynamic description of the heat transfer processes in the thermoelectric materials dedicated for the waste heat recovery. The link of these models with electrical properties of the material and a material solution based on a conducting polymer have also been presented in this paper.

1. Introduction

One of the main problems connected with the energy conversion processes taking place in the each branch of the industry (e.g. power engineering, chemical processing, metallurgy, transport, etc.) is the limited efficiency of the applied devices and machines (e.g. heat exchangers, pipelines, expanders, pumps, etc.) and inefficient organization of the energy conversion processes. Limited efficiency causes in the energy fluxes dissipation into the surroundings. In most of the cases these energy fluxes are not recovered, thus referred as the waste energy [1]. Waste energy can appear in the different forms such as heat, chemical (e.g. waste fuels), mechanical, pneumatic and the others. Many legal regulations related to the necessity of the increasing the energy generation efficiency [2], lowering the toxic and green gas emissions [3] resulted in focusing on the waste energy recovery as the chance for increasing the energy efficiency and lowering the consumption of the primary energy carriers [4]. Thus the waste energy recovery technologies became the object of the intensive scientific investigations in many of the scientific units worldwide. As it was described in [5] waste heat has the biggest share among the other waste energy sources, thus it is very interesting for the recovery. Depending on the heat source thermal properties and temperature, the waste heat sources can be divided into the high temperature (500—1200°C), medium temperature (250—500°C) and low temperature (80—250°C). Depending on the waste heat source parameters and power, many technologies can be used for the heat recovery [1]. It is possible to use waste heat directly for covering the energy demands of the energy conversion process in which the waste heat is generated (e.g. heating the substrates in the heat exchanger) or indirectly for the energy generation for other energy consumers not directly connected with this process (e.g. electricity and heat generation). From the thermodynamic point of view the most interesting waste heat recovery process is such, in which the low quality (low exergy) energy (such as waste heat) can be converted in the high quality (high exergy) energy (such as electricity). Since early 50-ties of the XX century the intensive research works are focused on technologies dedicated for the waste heat recovery into electricity [1]. Such processes are possible in energy systems based on the thermodynamic cycles. The main available technologies are external combustion engines (e.g. Stirling and Ericsson) [1], and steam power plants (e.g. Kalina and Organic Rankine Cycle). The detailed description of these systems, with the principle of their operation was comprehensively described in [1] and [6]. The main disadvantages of these systems is limited and relatively low efficiency (caused by irreversibility of the thermodynamic processes taking place in the cycle and limited efficiencies of the components) and high investment costs (as the consequence of the necessity of application of many expansive devices i.e. heat exchangers, pumps, working fluids, and the expanders). Thus, the application of other technologies, especially those based on the thermoelectric effect for waste heat recovery is very interesting. The thermoelectric materials are competitive when compared with the currently used technologies. Thermoelectric materials can be used for direct waste heat recovery from the heat source without the necessity of applying the expensive and low efficient heat engines. The conducting polymers are cheap and can be used in a wide range of geometries e.g. the bulk products, thin films, pristine materials, composites and the others. Thus, encouraged by the above described material properties, authors decided to prepare the experiment and analyse the process of the direct waste heat recovery via thermoelectric polymer together with the thermodynamic description of the heat exchange phenomenon in the material. In the following the description of the experiment and the mathematical models of the heat transfer in the material are presented.

2. Materials and methods

Basing on the literature survey [7], [8] and on the authors own experience the polyaniline was selected as the material that can potentially be interesting for the waste heat recovery. The selection was based on the polyaniline specific properties i.e. good thermal, environmental and chemical resistance and thermoelectric activity at the temperature below 500 K. Polyaniline has low density and can be easily formed in different shapes, what is important for the application. Authors decided to modify the polymer in order to obtain the material with the best possible thermoelectric properties. Thus the polyaniline was first protonated with the hydrochloric acid to obtain PANI-HCl(S) and then additionally filled with 20% wt. of nano-silver to create PANI-HCl(S)/Ag20(M). The materials were formed by press moulding into the rods with diameter of 10 mm and length of 80 mm. After the processing rods were mounted on the experimental setup. The test-stand was composed of the electric heater (simulating the waste heat source) and voltage and temperature measuring equipment. The experiment was carried and the generated voltage was measured. The more detailed description of the experiment and the materials properties were presented in [9].

3. Mathematical model of the heat transfer in the thermoelectric material and modelling results

The heat transfer model in the material was derived basing on the energy conservation equation and the Fourier's and Newton's law of the heat conduction in the material and the heat dissipation from the sample surface to the surroundings. This problem is commonly known as the heat exchange in a straight, non-insulated finite length rod. Because of the length and complexity of the derivation of the mathematical model the authors decided to present here only the final equations. For more details of the mathematical derivation of this model please refer to [10]. The temperature distribution in the material can be described by the relation

$$\Theta(x)=\Theta_0\cdot\frac{\cosh\left[m(L-x)\right]+P_L\cdot\sinh\left[m(L-x)\right]}{\cosh(mL)+P_L\cdot\sinh(mL)} \tag{1}$$

where x is the rod length coordinate, Θ(x) is the rod temperature excess in relation to the temperature of the surroundings across the rod length, Θ_0 is the rod temperature excess in relation to the temperature of the surroundings in restrain point, m is the rod temperature coefficient, P_L is the non-insulated surface coefficient, L is the total length of the rod.
Rod temperature coefficient is given by the relation

$$m=\sqrt{\frac{O\cdot h}{A\cdot k}} \tag{2}$$

where O is the rod perimeter, h is the surface film conductance at the side rod surfaces, A is the rod cross section area, k is the heat conductivity of the material.
P_L coefficient is given by the relation

$$P_L=\frac{h_L}{k\cdot m} \tag{3}$$

where h_L is the surface film conductance at the non-insulated rod end.
The total heat flux transferred in the rod can be expressed by the relation

$$\dot{Q}_{x=0}=\Theta_0\cdot\sqrt{A\cdot h\cdot O\cdot\lambda}\cdot\frac{tgh(mL)+P_L}{1+P_L\cdot tgh(mL)} \tag{4}$$

The Seebeck's voltage can be expressed by the relation

$$U_S=\alpha\cdot(T_2-T_1)=\alpha\cdot\Delta T \tag{5}$$

where α is the Seebeck coefficient, T_1 and T_2 is the temperature of the left and the right rod end face respectively.
The temperature difference on the both of the rod end faces can also be calculated for eq. (1) as follows

$$\Delta t=\Theta_0-\Theta_L=\Theta_0-\frac{\Theta_0}{\cosh(mL)+P_L\cdot\sinh(mL)}=\Theta_0\left(1-\frac{1}{\cosh(mL)+P_L\cdot\sinh(mL)}\right) \tag{6}$$

Combination of the eq. (5) and eq. (6) yields to the following relation

$$U_s=\alpha\cdot\Theta_0\left(1-\frac{1}{\cosh(mL)+P_L\cdot\sinh(mL)}\right) \tag{7}$$

Eq. 7 is the mathematical model of the heat transfer and voltage generation in the sample. Using this equation it is possible to model the generated voltage in rods made of different materials with different dimensions and in different operating conditions (surroundings conditions).

Described above mathematical model was applied in the computer software for the same thermoelectric material and the same heat exchange conditions as in the experiment. Two materials i.e. PANI-HCl(S) and PANI-HCl(S)/Ag20(M) were investigated. The analysis was performed for the following values of the parameters: d=10 mm, L=80 mm, Θ_0=180 °C, t_s=20 °C, h=9 W/m^{2}K, h_L=12 W/m^{2}K, $\alpha=2100\cdot10^{-6}$ V/K, k=0.18 W/mK for PANI-HCl(S) and $\alpha=3290\cdot10^{-6}$ V/K, k=0.22 W/mK for

PANI-HCl(S)/Ag20(M). The comparison between experimental results and modelled values are presented in the following.

Fig. 1 shows the temperature distribution (calculated from eq. 1) in both samples. The green broken line refers to the PANI-HCl(S) while the violet broken line refers to the PANI-HCl(S)/Ag20(M). For the comparison of the temperature distribution in investigated samples (which are characterized by relatively low thermal conductivities) the temperature distribution in the rod made of metal i.e. cooper (red solid line) and steel (blue dot line) with the same dimensions was also presented on this figure.

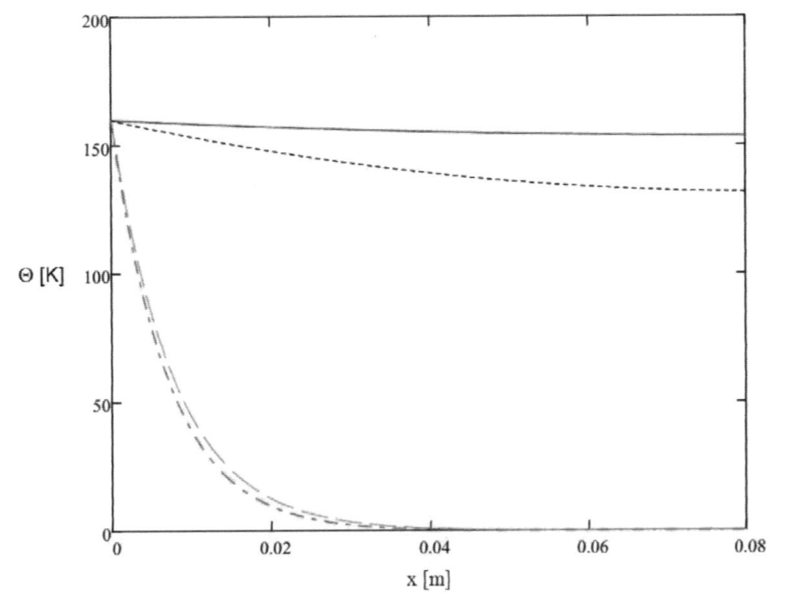

Figure 1. Temperature distribution in tested samples.
red solid line – rod made of cooper; blue dot line – rod made of steel;
green broken line – rod made of PANI-HCl(S);
violet broken line – rod made of PANI-HCl(S)/Ag20(M)

As it can be seen from the fig. 1 the decrease of the temperature in both of the tested thermoelectric materials is more rapid that in the case of cooper and steel. The minimal temperature is achieved in the sample centre. It is due to the material properties (lower thermal conductivity than in the case of metals). Such temperature distribution is advisable in the case of thermoelectric cooling as the temperature of the cooled surface should be comparable to the temperature of the surroundings. As it can be seen from the comparison of the temperature distribution in samples made of PANI-HCl(S) and PANI-HCl(S)/Ag20(M) the silver incorporation into the polymer results in more rapid temperature decrease. It is due to the increased thermal conductivity of this material.

The Seebeck's voltage was also calculated from the model (eq. 7) and compared with the experimental results. For PANI-HCl(S) the measured value was U_s=0.329 V while modelled U_{sm}=0.351 V. For PANI-HCl(S)/Ag20(M) the measured voltage was U_s=0.505 V while modelled U_{sm}=0.526 V. As it can be seen the silver addition into the polymer causes in the rise of the Seebeck's voltage. Moreover there is a good matching between the experimental and the modelled results.

IOP Conf. Series: Materials Science and Engineering **113** (2016) 012022 doi:10.1088/1757-899X/113/1/012022

4. Summary and conclusions

In this article the problems related to the waste heat recovery via selected thermoelectric materials were presented and analysed. The mathematical model of the heat transfer in the thermoelectric material, valid for rod samples, was developed and tested. This model can be applied for the simulations and modelling of the Seebeck's voltage changes in terms of different material properties i.e. thermal conductivity, sample dimensions and heat exchange conditions from the sample to the surroundings. Moreover it is possible to determine the heat flux transferred in the sample. The results of the experimental and modelling works showed that the thermoelectric materials can be interesting alternative when compared to the currently used waste heat recovery technologies. The further works in this topic should consider modelling of other geometries (e.g. ribs, thin surfaces, etc.) for designation of the optimal shape of the thermoelectric material for the waste heat recovery purposes.

5. References

[1] Szargut J 1993 *Industrial Waste Energy-Utilization and Equipment* (WNT: Warsaw)
[2] Collective work 2012 *Directive 2012/27/EU on energy efficiency* (European Parliament: Brussels)
[3] Collective work 2009 *Directive 2009/29/EC on the improve and extend the greenhouse gas emission allowance trading scheme of the Community* (European Parliament: Brussels)
[4] Minea V 2014 *Appl. Therm. Engineering* 69, **143-154**
[5] Kolasiński P 2015 *Energies* 8(5), **3351-3369**
[6] Bao J, Zhao L 2013 *Renew. Sustain. Energy Rev.*, 24, **325-342**
[7] Rogers S, Kaiser A 2004 *Current Applied Physics* **4** 407–410
[8] Wu J, Sun Y, Xu W, Zhang Q 2014 *Synthetic Metals* **189** 177–182
[9] Kolasińska E 2015 *Sterowanie strukturą polianiliny i jej kompozytów o właściwościach termoelektrycznych* (Electrotechnical Institute: Warsaw)
[10] Lienhard J IV and Lienhard J V 2015 *A Heat Transfer Textbook* (MIT: Massachusetts)

IOP Conf. Series: Materials Science and Engineering **113** (2016) 012023 doi:10.1088/1757-899X/113/1/012023

Polymer Materials for the Heat Recovery

E Kolasińska[1], P Kolasiński[2], B Mazurek[1]

[1]Electrotechnical Institute
Division of Electrotechnology and Materials Science, Wrocław, PL

[2]Department of Thermodynamics, Theory of Machines and Thermal Systems
Faculty of Mechanical and Power Engineering
Wrocław University of Technology, PL

E-mail: kolasinska@iel.wroc.pl

Abstract. Many of the processes in the industry, agriculture and microscale systems are associated with the waste heat generation, which often may be a menace or lower the efficiency of the processes. The thermoelectric cooling is becoming increasingly popular and gives the possibility to convert waste heat into electricity. The current thermoelectric cooling solutions are based on alloy materials. However, the new technologies pay attention to the environment burden, moreover the regulations of the production and recycling are becoming more and more restrictive. Conducting polymers are thermoelectrically active at low temperatures, cheap and environmentally safe. In this paper authors discuss the possibility of the application of conducting polymers for the heat recovery. Due to the operating temperature range and different nature of the waste heat sources, polymers might be an interesting solution and a complement for alloy-based thermoelectric materials. The character and nature of the formation of waste heat sources and conventional technologies of its recovery are also described in this paper. Moreover the advantages of thermoelectric cooling with the use of polymers are presented and two materials based on polyaniline are proposed.

1. Introduction

The energy conversion and technological processes in the industry, agriculture, motorization, domestic appliances, as well as microsystems are associated with the heat generation. In some of the cases the heat generation may be unsafe and may reduce the efficiency of the main process or negatively influence the operating conditions of machines or electronic equipment. Regardless of the phenomenon scale, the cooling is needed in most of the cases. The cooling methods based on the use of fans, radiators or refrigerating cycles are not always feasible, because of the freons toxicity, progressive miniaturization, specialization and complexity of the devices. The thermoelectric cooling, which also gives the possibility to convert the waste heat into the electricity, is becoming increasingly common.

2. The thermoelectric phenomena and parameters

In the conductors and semiconductors, the interdependent processes of electric charge and heat transport can occur. This means that the thermal process generates the charge transport, or vice versa, the flow of the electric current causes the thermal effect. These phenomena may occur in one conductive material or at the interface between two different materials. The thermoelectric Seebeck effect considered in this paper consist in the heat conversion into the electricity and it can be described mathematically by the eq. (1) [1, 2].

$$U_S = \alpha \cdot (T_1 - T_2) \tag{1}$$

where *Us* [V] – generated voltage, α [V/K] - Seebeck coefficient, T_1 - T_2 [K] - temperature gradient. Seebeck phenomenon can be applied in the energy storage devices and electrical generators.

The efficiency of the heat transfer and the energy conversion in the thermoelectric materials and devices can be estimated with the dimensionless thermoelectric figure of merit *ZT* (2).

$$ZT = \frac{\alpha^2 \sigma T}{\lambda} = \frac{\alpha^2 T}{\rho \lambda} \tag{2}$$

where α [V/K] - Seebeck coefficient , σ [S/m] - electrical conductivity, ρ [Ω·m] - electric resistivity (inverse of conductivity), λ [W/m·K] - thermal conductivity, T [K] - absolute temperature [3, 4]. Also, the value of the figure of merit *Z* with the dimension [K^{-1}] is popular in the literature and it is commonly used for the material efficiency estimations. For the practical application the thermoelectric materials with a *ZT* approx. 3 are needed. Such value is comparable to the efficiency of mechanical cooling and would allow to use the thermoelectric materials in a larger scale [1].

3. The alloy thermoelectric materials

The development of thermoelectric generators and coolers is observed since the 2nd half of the twentieth century. Currently, the most common thermoelectric materials are the modified semiconductors e.g. bismuth telluride Bi_2Te_3, bismuth antimonide BiSb or multicomponent alloys, which are produced in the high - temperature metallurgical processes. The value of *ZT* for this materials is ca. 1 i.e. in case of Bi_2Te_3/Bi_2Se_3 (25 mol%) $ZT = 0,6$ and in case of Bi_2Te_3/Sb_2Te_3 (75 mol%) $ZT = 0,9$ [1]. Fleurial et al [6] summarized the common alloy materials and shown that the most of these materials have the best thermoelectric efficiency at temperatures above 400 K.

Table 1. The properties of chosen semiconductors and metals [5, 6]

Material	σ [1/Ω·cm]	α [μV/K]	$\alpha^2\sigma$ [W/mK^2]	λ [W/mK]	Z [K^{-1}]
Bi_2Te_3	1000	200	$4,0\cdot10^{-3}$	1,6	$3,0\cdot10^{-3}$
PbTe	450	20	$2,6\cdot10^{-3}$	2,0	$1,2\cdot10^{-3}$
SiGe p-type	758	144	$1,6\cdot10^{-3}$	4,8	$3,3\cdot10^{-4}$
SiGe n-type	990	-136	$1,8\cdot10^{-3}$	4,45	$4,1\cdot10^{-4}$
Cu	580000	1,83	$1,9\cdot10^{-4}$	398	$4,8\cdot10^{-7}$
Ni	138889	-19,5	$5,3\cdot10^{-3}$	90,5	$5,9\cdot10^{-5}$
Ti	23810	9,1	$2,0\cdot10^{-4}$	21,9	$9,1\cdot10^{-6}$

The electrical conductivity (σ), the Seebeck coefficient (α), the thermoelectric power factor ($\alpha^2\sigma$), the thermal conductivity coefficient (λ) and the thermoelectric figure of merit (Z) for chosen metals and semiconductors are presented in Table 1. Metals have highest values of the electrical conductivity. However, the *ZT* (*Z*) of these materials are even 3-4 orders lower than in the case of the alloy materials. Semiconductors are characterized by much higher values of Seebeck coefficient, which affects the value of *Z* in a second power (eq. 2). Their thermal conductivity is significantly lower than in the case of metals. Such parameters, despite the high electrical conductivity, causes that the thermoelectric figure of merit *Z* for metals is in the range 10^{-8} – 10^{-5} [K^{-1}]. Raw semiconducting alloys are characterized by the values of the thermoelectric figure of merit *Z* from the range of 10^{-5} – 10^{-3} and the suitable modification can improve it.

4. The polymer thermoelectric materials

The technologies of the 21st century pay attention to the environment burden and the regulations in this field are becoming more and more restrictive. The recovery and reuse of raw materials are also nowadays very important. Alloy inorganic materials burden the environment through the life cycle of the ore, refining processes and scrap materials. The synthesis and molding processes are high energy consuming thus, the costs of the products are high, but the recovery and recycling of waste are negligible. Environmentally friendly technology should eliminate or at least limit the use of heavy metals and reduce the energy consumption of the production processes.

Conductive polymers can be a good complement to the classic alloy materials, reducing the environmental burden due to their chemical composition, lower manufacturing and processing costs and the recycling ability. Polymers are lightweight, can be formed in variety of shapes [5, 8, 9] and are thermoelectrically active at the ambient temperature. It is still a relatively new group of materials so even a slight change in their properties may become a breakthrough in the field of the waste heat conversion, removal and recovery [10]. The modified conductive polymers may have the value of the thermoelectric power factor ($\alpha^2\sigma$) from 10^{-8} up to 10^{-3} W/mK^2 and Seebeck coefficient from several to hundreds μV/K. These properties are listed in Table 2 for the selected materials In the case of these materials the properties are comparable to the alloy materials. The maximum value of the conductivity of polyacethylene reaches 10^6 S/cm, which is comparable to metals. Lower values are observed for other polymers [11]. This shows how the different modification methods effect the polymer and how important is the choice of the modifier.

Table 2. The properties of the selected conductive polymers [5, 7, 12, 13]

Polymer	Modifier	σ [(Ω·cm)$^{-1}$]	α [μV/K]	$\alpha^2\sigma$ [W/mK^2]
Polyacethylene (PAC)	-	6405	20,6	$2{,}7\cdot10^{-4}$
	I*	60000	15	$1{,}3\cdot10^{-3}$
Polyaniline (PANI)	-	18	3	$1{,}6\cdot10^{-8}$
	CSA*	200	10	$2{,}0\cdot10^{-6}$
Polypyrrole (PPY)	-	26	5	$6{,}5\cdot10^{-8}$
	PANI	15	7	$7{,}4\cdot10^{-8}$

*I – iodine
**CSA – camphor sulfonic acid

5. Experimental

Polyaniline was chosen by the authors because of its properties typical for polymers i.e. good heat resistance, environmental and chemical stability. PANI can be compounded with other materials common in the industry (e.g. metals, glass, plastics) in the form of composites and layer systems. The main problem in the practical application is improving its thermoelectric efficiency. The modification with the use of the protonic acids or conducting additives is referred to be the solution to improve the thermoelectric efficiency [14].

The authors obtained products based on polyniline with the parameters shown in Table 3. Both materials were obtained by chemical oxidative method at a temperature of 0-5 °C. The synthesis method was comprehensively described in [15]. The PANI-HCl material is the polyaniline protonated with hydrochloric acid, the PANI-HCl-Ag is its composite filled with 20% wt. of silver. The raw materials: aniline 95% (Chempur), hydrochloric acid 35 – 37% (Chempur) and silver nanopowder <100 nm 99,5% (Sigma Aldrich) were used as purchased.

The measurement of Seebeck voltage, which determines the Seebeck coefficient α in accordance with equation (1) was carried out with a temperature gradient of 300 - 453 K using a digital multimeter HP34401A. The degradation temperature *DT* was determined by thermal analysis performed on TGA/DSC1 Mettler Toledo. Thermal conductivity λ was measured by Poensgen method and the resistivity ρ_v using Keithley 6517. The value of the thermoelectric figure of merit *ZT* at 453 K was determined according to equation (2). The materials should be applied at temperatures below 550 K. Their efficiency *ZT* is similar to that of the unmodified alloys, and the Seebeck coefficient is significantly greater.

Table 3. The properties of polyaniline based materials

Material	ρ_v [Ω·cm]	λ [W/mK]	α [μV/K]	*ZT*	DT [K]
PANI-HCl	19,57	0,18	2193,55	$6{,}19\cdot10^{-2}$	573
PANI-HCl-Ag20	16,43	0,22	3290,32	$1{,}36\cdot10^{-1}$	598

6. The low-temperature waste heat sources and the standard methods of the waste heat recovery

The low-temperature waste heat sources may have different physical form and geometry, as well as chemical nature and corrosivity. The thermal and power output characteristics of such heat sources may also be various, thus waste heat can be generated at the different temperature levels and different thermal power. The waste heat is generated in almost all of the energy conversion processes, especially in the following branches of the industry and households [16]:

- *power engineering* (exhaust gases 250 - 1200°C, coolants 40 - 150°C, solid and liquid waste 40 - 200°C, hot surfaces 40 - 400°C, waste steam 150 - 350°C),
- *chemical industry* (gases 100 - 600°C, liquids 40 - 200°C, hot surfaces 40 - 300°C),
- *food industry* (liquids 40 - 100°C, cooling air 50 - 100°C, hot surfaces 40 - 400°C),
- *metallurgy* (exhaust gas 250 - 1200°C, process gases 300 - 1400°C, coolants 40 - 150°C, solid and liquid waste 40 - 200°C, hot surfaces 40 - 700°C),
- *motorization* (exhaust gases 500 - 1100°C, coolants 40 - 100°C, hot surfaces 40 - 500°C),
- *households* (flue gases 150 - 300°C, liquids 40 - 90°C).

The main available technologies for waste heat recovery from such sources are external combustion engines [16], and steam power plants (e.g. Kalina and Organic Rankine Cycle). The detailed description of these systems, with principle of their operation was comprehensively described in [16] and [17]. The main disadvantages of these systems is limited and relatively low efficiency and high investment costs. Thermoelectric material can be used for direct waste heat recovery from the heat source without the necessity of applying the expensive and low efficient heat engines. Thus, the thermoelectric materials are very interesting and perspective for the waste heat recovery when compared with the currently used technologies.

7. Summary and conclusions

In this paper the comparison of the different thermoelectric materials possible for application in the waste heat recovery and cooling of devices was presented. In particular the alloy-based and the polymer thermoelectric materials were analysed. Moreover the results of the experimental analyses of the specially designed thermoelectric polymers were here presented. The results of the comparison and experimental works proves the following conclusions:

- The composites based on polyaniline can be a complement for alloy materials at lower temperatures,
- The thermoelectric parameters of thermoelectric polymers are promising,
- These materials may be formed in variety of shapes (coatings, bulk products) and compounded with other materials (glass, plastics, metals), what is the advantage in comparison with the currently used waste heat recovery technologies.
- The discussed examples of the waste heat sources show the proper conditions (temperature range and gradient) for the application of the materials based on polyaniline.

8. References

[1] Królicki Z 2006 *OWPWr* 11-57
[2] Toshima N, Yan N, Kajita M 2002 *21st Int. Conf. on Thermoelectronics IEEE* 147-150
[3] Hostler S, Kaul P, Qu V 2006 *ITHERM* 1400-1405
[4] Li J, Tang X, Li H, Yan Y, Zhang Q 2010 *Synthetic Metals* **160** 1153–1158
[5] Shakouri A, Li S 1999 *18th Int. Conf. on Thermaelectronics IEEE* 402- 406
[6] Fleurial J, Borschevsky A, Caillant T 2000 *ACS JPL Technical Report Server* 1080-1085
[7] Park Y 1991 *Synthetic Metals* **45** 173-182
[8] Chatterjeea K, Gangulyb S, Karguptac K, Banerjeea D 2011 *Synthetic Metals* **161** 275–279
[9] Yan H, Ohta T, Toshima N 2001 *Macromolecular Materials and Engineering* **286** 139–142
[10] Ivanov S, Tsakova V 2005 *Electrochim. Acta* **50** 5616-5623
[11] Kaiser A 2001 *Adv. Mater.* **12-13** 927-941
[12] Holland E, Monkman A 1995 *Synthetic metals* **74** 75-79
[13] Salaneck W, Bredas J 1994 *Solid State Communications* **92** 31-36
[14] Li J, Tang X, Li H, Yan Y, Zhang Q 2010 *Synthetic Metals* **160** 1153–1158
[15] Kolasinska E., Mazurek B 2014 *IAPGOS* **4** 94-97
[16] Szargut J 1993 *Industrial Waste Energy-Utilization and Equipment* (WNT: Warsaw)
[17] Bao J, Zhao L 2013 *Renew. Sustain. Energy Rev.*, **24**, 325-342

Energy harvesting in high voltage measuring techniques

Paweł Żyłka[1] **and Marcin Doliński**[2]
[1] Department of Electrical Engineering Fundamentals, Wrocław University of Technology, Wyb. Wyspiańskiego 27, 50-370 Wrocław, Poland
[2] Accenture Sp. z.o.o., ul. Nyska 83/85, 50-505 Wrocław, Poland

e-mail: pawel.zylka@pwr.edu.pl, marcin.jakub.dolinski@gmail.com

Abstract. The paper discusses selected problems related to application of energy harvesting (that is, generating electricity from surplus energy present in the environment) to supply autonomous ultra-low-power measurement systems applicable in high voltage engineering. As a practical example of such implementation a laboratory model of a remote temperature sensor is presented, which is self-powered by heat generated in a current-carrying busbar in HV-switchgear. Presented system exploits a thermoelectric harvester based on a passively cooled Peltier module supplying micro-power low-voltage dc-dc converter driving energy-efficient temperature sensor, microcontroller and a fibre-optic transmitter. Performance of the model in laboratory simulated conditions are presented and discussed.

1. Introduction

Endless supply of energy and perpetual power has been a pipe dream of philosophers and engineers manifested in *perpetuum mobile* conception. Perpetual motion machine working indefinitely without any energy source cannot be unfortunately made as it violates the fundamental thermodynamic laws of contemporary physics. However, electronic instruments which live on extracting energy from seemingly perpetual environmental sources and converting it into electricity do actually already exist.

Energy, in various forms – thermal, light, mechanical, electromagnetic (e-m) radiation – is present in our direct surrounding. However, depending on the source, energy density and instantaneous power may be too small making direct supply of technical appliances impossible, especially after additional conversion to electrical energy. Energy harvesting (termed also as *energy scavenging* or *power harvesting*) is an innovative scientific and industrial trend, relaying on extracting environmental energy in small chunks from the direct neighbourhood of the appliance, converting it into electricity and storing it for a discontinuous use. The concept behind this approach is to secure energy self-sufficiency for ultra-low power microelectronic appliances and sensor devices, working in intermittent mode of operation without access to a traditional power network or any kind of replaceable storage battery. Alternative ambient energy sources used in power harvesting are either renewable (like outdoor light, air or water movement) or are related to manmade derivative power sources, from which a surplus energy, dissipated into the environment, is drawn (like indoor artificial light, exhaust heat, mechanical vibrations, electric and magnetic fields, forced fluid and air flow, e-m waves etc.). Selection of a particular environmental energy source is thus dictated by its accessibility in the surrounding area of a supplied apparatus, energy conversion strategy and a required power density.

Energy scavenging provides thus access to energy available locally in practically unrestrained amount but at low or very low momentary power and makes devices totally autonomous, in terms of control and supply. When combined with short range, low power wireless communication and mesh networking it allows to give up structural and power electric wiring, makes sensor systems more flexible and fault tolerant, enhances operation safety, and makes logistics much simpler. Autonomous sensors fitted with power harvesters hardly require servicing, which is a valuable feature especially in monitoring of areas or processes to which access cannot be restricted or which cannot be stopped. Battery-less supply is thus predisposed for electronic sensor systems installed in remote, difficult to reach or hazardous locations, where other categories of electrical supply are inviable or unmanageable. Current trend in construction of energy harvesters is to utilize purely "material" energy conversion by making use of physical phenomena, which yield potential difference. In many cases such generated voltage not directly available for electronics supply or a source-load impedance mismatch is present thus a converter is another vital part of almost every energy scavenger. And as energy accessible from environmental sources is recurring, instable or at low power a harvester must also be equipped with a kind of energy storage arrangement. It must be thus emphasized that energy harvesting does not assert non-interruptible or robust energy supply and therefore, electrical devices supplied by this method must be designed and adapted for irregular operation with unpredictably long out-of-supply intervals. Harvester-supplied device operates thus totally randomly and starts functioning only when the amount of accumulated electrical energy is sufficient for commencement and accomplishment of its task and then it goes into deep sleep state, according to a recurring scheme snooze-on-snooze.

Energy harvesting is currently an excitable subject matter in science and engineering. Extensive reviews on energy harvesting and related topics have been already presented by many authors: Matiko *at al.* discussed current trends in application of energy harvesting in buildings [1], Harne and Wang examined current research in vibration energy harvesting using bistable systems [2], Davidson and Mo considered energy harvesting schemes in structural health monitoring [3], Elvin and Erturk collected advances in energy harvesting methods [4] while Tan presented efforts to design, analyze and implement energy harvesting in autonomous sensor systems [5] – just to name a few.

Energy harvesting is also gradually finding its application in high voltage (HV) engineering, although such efforts are not so numerous as it may appear feasible. Energy harvesting seems to be especially suitable for supplying sensors and measuring devices installed in HV-biased locations, where access and servicing is limited only to occasional situations when HV is off. Zangl *at al.* presented an elegant approach to construct autonomous on-line monitoring network for HV overhead power lines in which sensor nodes, mounted directly on HV conductor cables were supplied by capacitive dividers deriving power from the electric field provided by the HV line itself. At 150 kV 50 Hz AC such power harvesting nodes were capable of delivering approximately 370 mW [6]. Guo *et al.* have extended Zangl's idea by simulating performance of the capacitive harvesting device up to 270 kV and also proposed a linear harvester design intended for HV DC lines, relaying on small winged rectangular winding structure propelled by wind to swirl around the HV conductor line and produce electric power due to electro-magnetic induction in the static magnetic field emanated from the HV DC line. A hypothetical, simulated average power output of such device at 2000 A and 7 m/s wind speed was 1 mW, which was enough to trigger XBee RF wireless radio every 15 min [7]. Zhue *et al.* simulated energy harvesting from electric filed in HV AC substations by means of a large electrode located close to the ground and electromechanical synchronous switching circuit [8]. Roscoe and Judd experimented with harvesting power from ambient AC magnetic field near safely accessible HV appliances (transformers, post insulators, disconnectors, wall bushings, or circuit breakers) located in HV substations, showing that 300 μW may be harvested from 50 Hz magnetic flux density of 18 μT [9]. mNODE (Micropelt) is a commercially available remote temperature sensor powered by inductive energy harvester, converting stray magnetic fields nearby electrical infrastructure elements (like current busbars or power distribution arrangements) into electricity. mNODE is claimed to start its operation at 50 A, is sized 40×40×26 mm and transmits data using IEEE802.15.4 compliant radio [10]. Electricity may be also produced by harvesting surplus heat dissipated in HV appliances. Olekshii *et al.* investigated energy recovery from thermal loses generated by HV power transformers by means of thermoelectric modules. Power produced this way was insignificant comparing to the losses and power rating of the transformer and thus it may not be considered a mean to enhance energy efficiency of

power transformers, it may however energize low-power sensors [11]. Drawing power from the heat radiated by the current-carrying busbar is another example of harvesting-oriented approach in HV engineering. Such solution using remote TE-qNODE sensor node supplied by a harvester based on advanced thin-film thermoelectric module has been jointly presented by Micropelt and Schneider companies. Unfortunately, as a commercial construction, it was lacking a comprehensive technical data. The only information made available was the intended application to low- and medium-voltage switchgear, minimal temperature difference between a busbar and surrounding air of 5 K and use of radio link to convey data at 60 s interval between consecutive temperature measurements [12].
The presented paper aimed thus at investigating the feasibility of using a harvester based on regular thermoelectric Peltier module to supply low-power microprocessor-based sensor node to measure temperature of MV armature (like a busbar). The goal of the work was also to explore a practicability of fiber optic link as a mean to transmit measurement data in order to make the harvester-supplied node totally immune to electromagnetic noise and capable of working in enclosed three-phase electric power systems without mutual interferences.

2. Apparatus

A laboratory harvester model was build using a commercially available mid-range thermoelectric module TEC-18351 (I_{max}=3.9 A, U_{max}=15.5 V, P_{max}=34 W, R_{int}=3.5 Ω, 25x25x3.5 mm). The module was fitted with a slotted 9-fin 38x38x20 mm aluminium radiator (thermal resistance unknown), permanently affixed to the module plate with a thermally conductive glue. The radiator-module assembly was placed on a 50 W copper hot plate, simulating a real heat source. The temperature of the hot plate was stabilized using YUDIAN AI-208 PID controller and a Pt100 temperature sensor. Voltage and current output of the thermoelectric module as well as temperature of the heater and the radiator was monitored using a PC-interfaced Agilent 34972/34901A data acquisition unit.

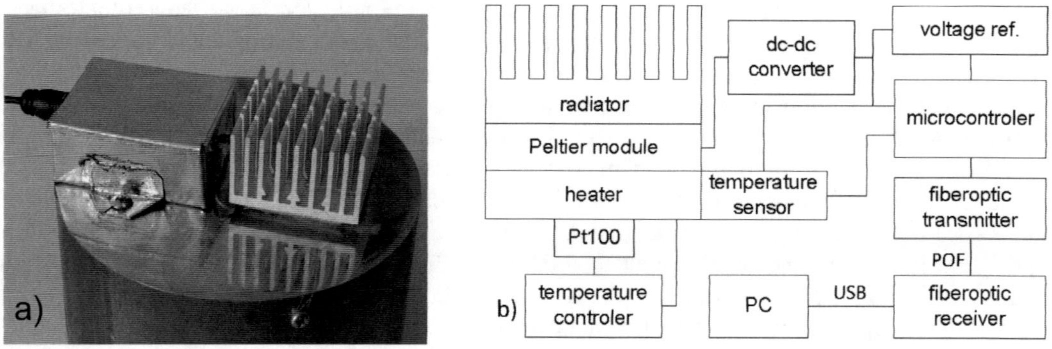

Figure 1. Test set-up: a) thermoelectric harvester and sensor node, b) schematic diagram.

In order to boost a mV voltage output of the thermoelectric module to a level suitable for supplying microelectronic devices an ultralow voltage step-up dc-dc converted was exploited, build using 1:100 step-up transformer (LPR6235-752SML, Coilcraft) and LTC3108 chip (Linear Technology) [13]. Its output voltage was programmed to a minimal permissible value (2.35V). The auxiliary output of the converter (to which sensor node supply terminal was connected) was energized by PGD signal i.e. only when voltage across the main output storage capacitor (470 μF, replaced by 330 μF in the final version of the harvester) was within operational range. Such mode of operation resulted in hysteretic switching the sensor node on for a short interval (when the voltage arrived at 92.5% of the programmed value) and then turning it off (when the voltage dropped below 91 % of the programmed value) for a longer time period until the voltage again reached 92.5% of the programmed value.
A model sensor node was build using ATmega328p (Atmel) microcontroller (μC), clocked at 8 MHz, fitted with MAX6605 precision, low power analog output temperature sensor. Measurement data

transfer was accomplished using IR LED diode directly driven at 38400 bps by μC USART transmit pin through 330 Ω resistor. The optical power of the LED was channeled to 1 mm in diameter plastic optical fiber (POF) using a modified fixture (FC684208T, Cliff). All PCB-mounted electronic elements were shielded in a small metal enclosure, from which only POF cable and Peltier module electrical connectors were guided, as it was shown in figure 1a.

A fiber optic receiver (custom-build according to a modified Maxim An1117 reference design [14] and using IFD91 PIN photodiode mounted in a modified FC684208T fixture) was used to accept optical signals from the sensor node and translate them into USB datagrams using UART-USB interface chip FT232R (FTDI Ltd.). Measurement data was then recorded using an ordinary PC and a terminal program. A schematic diagram of the entire measurement set-up was shown in Figure 1b.

High voltage laboratory tests, simulating operation in HV conditions were carried out using a large bronze block, heated to 45°C, on which the harvester and sensor node were mounted (as in Figure 1a). The copper block was HV-biased using HV power amplifier (model 20/20C, Trek Inc.) controlled by a function generator (33522A, Agilent) supplying thus 0-20 kV DC or 50 Hz AC signals.

3. Experimental results and discussion

First of all, open circuit voltage and short current (across 10 Ω current shunt of Agilent 34901A multiplexer) generated by the radiator-thermoelectric module test assembly was evaluated. The assembly was oriented vertically (in order to endorse natural air convection) and placed into a good thermal contact with a test heat source, which temperature was set to 32, 41 and 50 °C consecutively. All those experiments were carried out in still air at ambient temperature 26 °C.

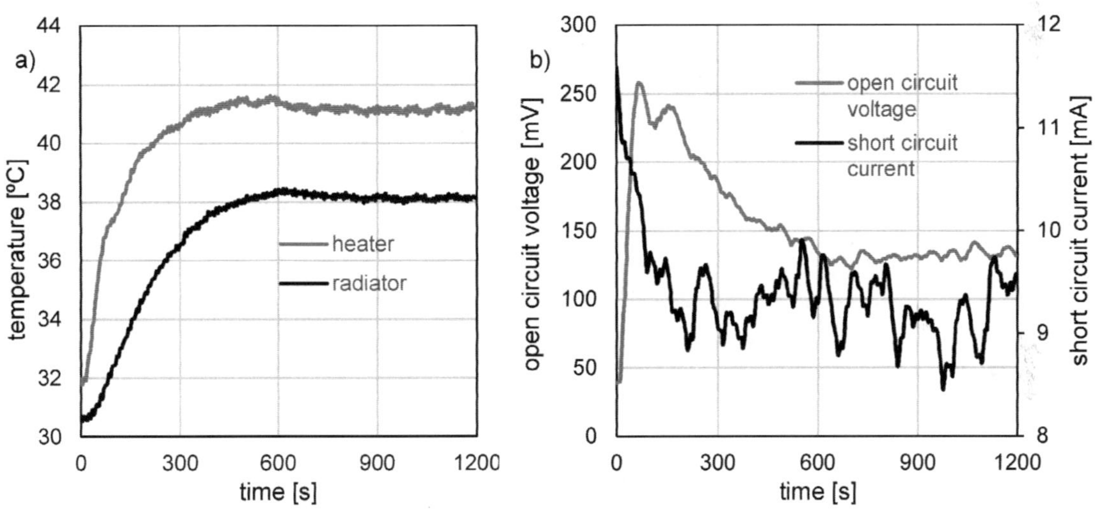

Figure 2. Thermoelectric module test assembly: a) heater and radiator temperature, b) open circuit voltage and current over 10 Ω current shunt.

Exemplary temperature, voltage and current curves recorded for the case of 41 °C were shown in figure 2 while table 1 lists mean output values obtained in stable thermal conditions. It should be pointed out that in still air at 26 °C ambient temperature and 32.5 °C heater temperature (thus at the temperature difference below 7 °C) approx. 0.15 mW was released at 10 Ω load. Moreover, it should be indicated that 10 Ω test load was not strictly matched with the internal resistance of the thermoelectric module therefore truly available power was slightly underestimated.

Table 2. Radiator-thermoelectric module test assembly output (mean values in stable thermal conditions, ambient temperature 26°C, sill air).

heater temperature (°C)	heater-radiator temperature difference (°C)	open circuit voltage (mV)	current at 10 Ω load (mA)	power at 10 Ω load (mW)
32	1.4 - 1.6	50 - 55	3.5 – 4	0.1 - 0.2
41	3.0 – 3.2	130 - 140	8.5 – 9.5	0.7 – 0.9
50	4.5 – 5.0	230 - 250	15.5 – 16.5	2.4 – 2.7

Voltage and power output of the thermoelectric assembly was thus far too low to be directly used to supply the sensor node and for its continuous operation. Therefore a boost dc-dc converter capable of operating at mV input voltages was employed together with a capacitive energy storage. The input impedance of the constructed converter was variable, as it may be determined in figure 3a and it was dependent on the input voltage. The impedance changed from approx. 12 Ω at 20 mV to approx. 6.5 Ω at 60 mV, then peaked to approx. 11 Ω at 100 mV and then decreased back to approx. 5.5 Ω at higher voltages, thus it was also not accurately matched with the (catalogue) internal resistance of the thermoelectric module making power transfer from the module to the load not ideal.

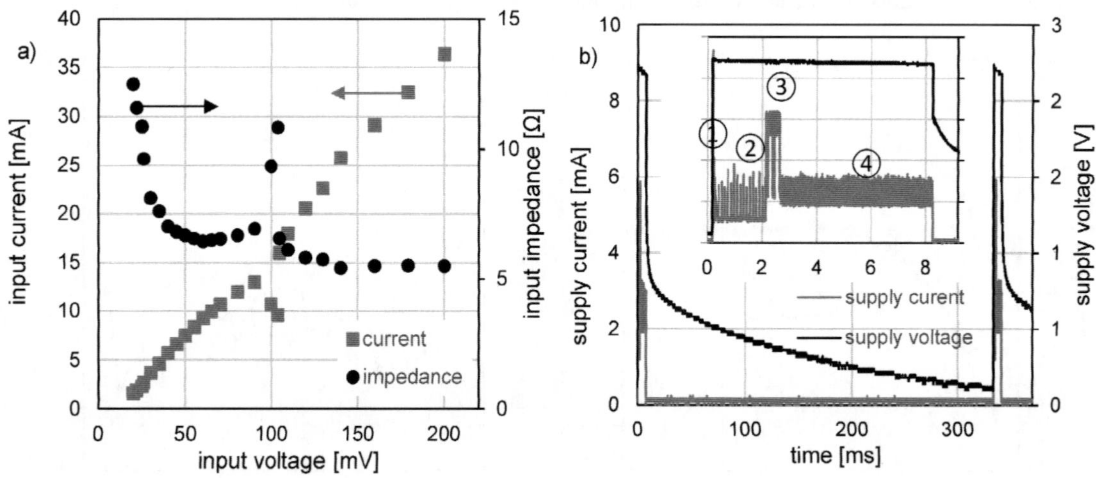

Figure 3. Supply of the sensor node: a) variable input impedance of the dc-dc converter, b) voltage and current trace during 2 on-off cycles.

The sensor node operation was made simple to minimize energy consumption. It was controlled by the firmware code, executed by the μC and the dc-dc converter cyclic operation. After μC power up and initialization (period marked (1) in figure 3b inset) 16 successive 10-bit analogue-to-digital conversions were performed (2) and their 14 bit accumulated result was finally sent out serially using USART and IR LED (3). After that μC operated in an empty loop until accumulated energy was consumed resulting in voltage drop below 91 % of the programmed value and thus sensor node de-energization (4). Such on-off cycle was repeated as soon as the energy transferred from the thermoelectric module to the storage capacitor resulted in voltage rise back to 92.5 % of the programmed value. The supply voltage and current drawn by the sensor node during such recurrent operation was shown in figure 3b. The mean supply current was approx. 2.3 mA while the maximal momentary value (during start-up) was approaching 9 mA; data transfer required approx. 6 mA.

The sensor node was therefore totally dispatched for a period of energy accumulation and resurrected each time enough energy has been collected. The time required by the dc-dc converter to recharge the storage capacitance back to the operational voltage (i.e. the time period between two consecutive run sequences of the μC) was dependent on its input voltage as it was shown in figure 4a. It should be noted that data presented in figure 4a was collected when supplying the node from a laboratory power supply of negligible internal resistance and current capability much higher than the node input current. Input voltage of 26 mV resulted in 0.18 runs/s repetition rate while at voltages exceeding 100 mV more than 5 runs/s were completed. Repetition rate 0.8-1.2 runs/s may be thus expected at 32 °C when taking into account data in table 1.

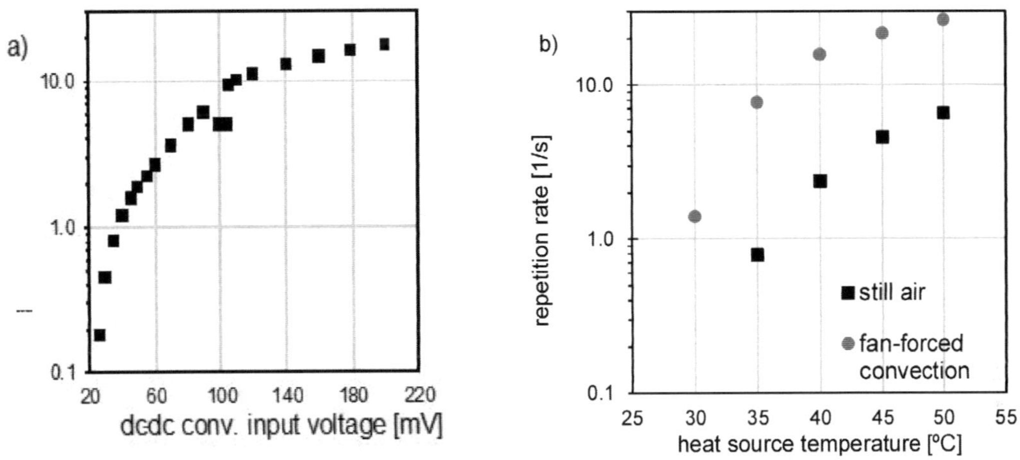

Figure 4. Sensor node on-off repetition rate: a) idealized supply case (laboratory power supply of negligible internal resistance), b) supplied by experimental thermoelectric harvester assembly.

Repetition rate measured when the sensor node was supplied exclusively by the thermoelectric harvester test assembly at 25 °C ambient air temperature was shown in figure 4b. The minimal heat source temperature necessary to make the sensor node active was 35 °C in still air (i.e. at 10 °C temperature difference). The minimal operational temperature was lowered to 30 °C (i.e. at just 5 °C temperature difference) when cooling of the radiator was improved by forced air movement, produced by a miniature 0.9 W electric fan positioned 1 m away from the thermoelectric module assembly.

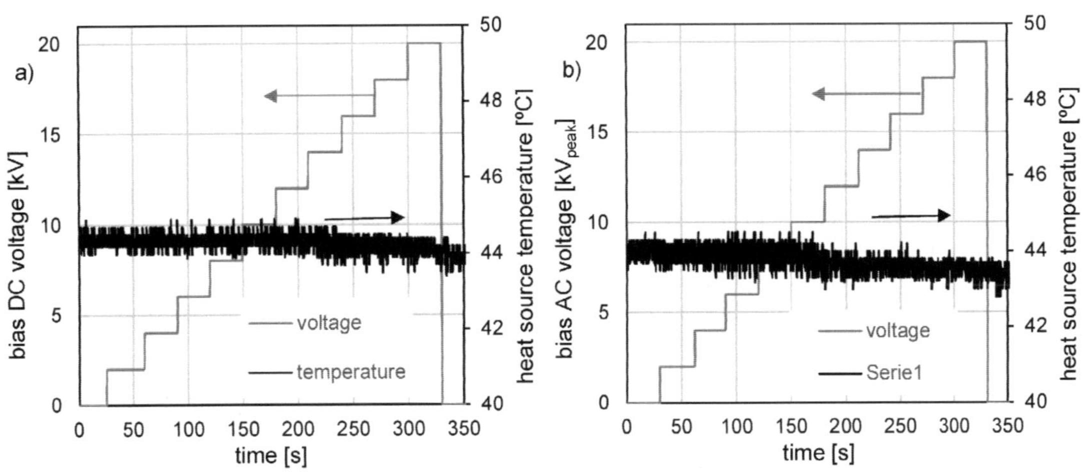

Figure 5. HV-biased sensor node operation: a) 0-20 kV DC, b) 0-20 kV 50 Hz AC.

Finally the sensor node operation was verified using a HV-biased heat source (heated to 45 °C and then allowed to naturally cool down). The HV bias was risen from 0 to 20 kV (DC or 50 Hz AC peak value) in 2 kV steps every 30 s and the heat source temperature was monitored by the senor node supplied only from the thermoelectric harvester assembly. Temperature measurement process was not affected by HV step rise as it was evidenced by the temperature records shown in figure 5. The uncertainty type A (expressed at 95 % confidence level) of the temperature measurement performed in the whole bias voltage range 0-20 kV was not worse than 0.9 °C; it was to some extent higher for HV AC bias (0.9 °C) than for DC (0.7 °C) but such difference may be considered insignificant.

4. Conclusion

It was shown that a commercial mid-sized thermoelectric Peltier module passively cooled with a small radiator was capable of generating electrical energy which boosted in voltage and capacitively accumulated may be used to power up low-power microcontroller-based sensor node at sub-second intervals. Model sensor node fitted with such energy harvesting assembly was operative at temperature difference between the heat source and ambient, still air of at least 10 °C while just 5 °C was sufficient in augmented cooling conditions. Infrared fiber optic data transfer was also a technically feasible solution in case of such low-power supply approach making the sensor node unaffected by electromagnetic interferences. Properly shielded and thermoelectrically-supplied sensor node may be safely operated under biased conditions up to 20 kV DC and 50 Hz AC making such solution useful in remote monitoring of armature in MV switchgear.

5. References

[1] Matiko J W, Grabham N J, Beeby S P and M J Tudor 2014 Review of the application of energy harvesting in buildings *Meas. Sci. Technol.* **25** 012002

[2] Harne R L and Wang K W 2013 A review of the recent research on vibration energy harvesting via bistable systems *Smart Mater. Struct.* **22** 023001

[3] Davidson J and Mo Ch 2014 Recent advances in energy harvesting technologies for structural health monitoring applications *Smart Mater. Res.* 2014 ID 410316

[4] ed Elvin N and Erturk A 2013 *Advances in energy harvesting methods* (New York: Springer Science + Business Media)

[5] ed Tan Y K 2013 *Energy harvesting autonomous sensor systems: design, analysis, and practical implementation* (Boca Raton: CRC Press Taylor & Francis Group)

[6] Zangl H, Bretterklieber T and Brasseur G 2009 A feasibility study on autonomous online condition monitoring of high-voltage overhead power lines *IEEE Trans. Instrum. Meas.* **58** 1789–96

[7] Guo F, Hayat H and Wang J 2011 Energy harvesting devices for high voltage transmission line monitoring *Proc. IEEE 2011 Power and Energy Society General Meeting* 24-29 July 2011, San Diego, CA USA

[8] Zhu M, Judd M D and Moore P J 2009 Energy harvesting in substations for powering autonomous sensors *Proc. SENSORCOMM 3rd Int. Conf. Sensor Tech. and App.* Athens Glyfada 2009, 246–51

[9] Roscoe N M and Judd M D 2013 Harvesting energy from magnetic fields to power condition monitoring sensors *IEEE Sens. J.* **13** 2263–70

[10] http://micropelt.com/mnode.php, hyperlink valid on 17.10.2015

[11] Olekshii R, Kukla N and Linkevics O 2014 Feasibility of usage of thermoelectric modules for recovering of low-potential heat from a surface of power transformers *RTUCON 55th Intl. Scientific Conf. on Power and Electrical Engineering,* Riga Technical University 124–9

[12] http://www.micropelt.com/applications/te_qnode.php, hyperlink valid on 17.10.2015

[13] LTC3108 Ultralow voltage step-up converter and power manager Linear Technology Corporation datasheet LT 0813 rev. C
[14] Small photodiode receiver handles fiber-optic data rates to 800kbps 2001 Maxim Integrated Products Application Note 1117

Institute of Physics
Dirac House, Temple Back
Bristol BS1 6BE UK

ISSN: 1757-8981
ISBN 978-1-5108-2229-0

CURRAN ASSOCIATES INC.
proceedings
.com

Joint Varenna-Lausanne International Workshop on the Theory of Fusion Plasmas 2016

Journal of Physics: Conference Series Volume 775

Varenna, Italy
29 August – 2 September 2016

IOP Institute of Physics